GLOBAL CLIMATE CHANGE

GLOBAL CLIMATE CHANGE

HORACE M. KARLING
EDITOR

Nova Science Publishers, Inc.
Huntington, New York

Senior Editors: Susan Boriotti and Donna Dennis
Editorial Coordinator: Tatiana Shohov
Office Manager: Annette Hellinger
Graphics: Wanda Serrano
Book Production: Matthew Kozlowski, Jonathan Rose and Jennifer Vogt
Circulation: Cathy DeGregory, Ave Maria Gonzalez, Ron Hedges, Mike Hedges, Andre Tillman

Library of Congress Cataloging-in-Publication Data
Available Upon Request.

ISBN 1-56072-999-6

Copyright © 2001 by Nova Science Publishers, Inc.

JK

         227 Main Street, Suite 100
         Huntington, New York 11743
         Tele. 631-424-6682   Fax 631-425-5933
         e-mail: Novascience@earthlink.net
         Web Site: http://www.nexusworld.com/nova

Printed in the United States of America

CONTENTS

LIST OF TABLES

LIST OF FIGURES

PREFACE

The Bush Administration recently announced that global warming may be imaginary and that the Kyoto Protocols are not acceptable and would be opposed, raising the ire of most developed countries. The Administration's own scientific commission followed with a report confirming the validity of the scientific basis for global warming. Since the United States Senate was not expected to ratify the Protocols in any event, the Bush Administration has reaffirmed its position that they were economically unjust to America and likely to result in negative impacts on the U.S. economy.

The likely scenario is that the United States government will probably either take some ameliorative actions or appear to do so, based on whether the current position can be politically massaged within the next twelve to eighteen months. Serious efforts on any of the underlying causes of global climate change are unlikely, based on the razor-thin election of George W. Bush blended with a heavy dosage of financial supporters from the energy industries who would be required to clean up their operations.

This much-needed book presents analyses of energy efficiency, the scientific research on global warming, the Kyoto Protocols, reduction techniques for greenhouse gases and carbon emissions, and policy options.

Chapter 1

GLOBAL CLIMATE CHANGE

Wayne A. Morrissey and John R. Justus

INTRODUCTION

There is concern that human activities are affecting the heat/energy-exchange balance between Earth, the atmosphere, and space, and inducing global climate change. Human activities, particularly the burning of fossil fuels, have increased atmospheric carbon dioxide (CO_2) and other trace greenhouse gases. If these gases continue to accumulate in the atmosphere at current rates; many believe global warming would occur through intensification of Earth's natural heat-trapping "greenhouse effect." Possible impacts might be seen as both positive and negative.

A warmer climate would probably have are reaching effects on agriculture and forestry, managed and un-managed ecosystems, including natural habitats, human health, water resources, and sea level depending on climate responses. Regional agricultural practices could change, yield stabilities might decrease in some regions, and survival over winter of some insect pests might increase. Forest productivity might decline in some regions; and changes in climate, when added to other environmental stressors, could produce major regional disturbances. Some climate modification, e.g., in northernmost growing regions, is thought to be beneficial for agriculture however.

Although casual relationships between projected long-range global climate trends and record-setting warmth and severe weather events of the past two decades have not been firmly established, attention has been focused on possible extremes of climate change and the need for better understanding of climate processes to improve climate model forecasts.

The basic policy question is: Given scientific uncertainties about the magnitude, timing, rate, and regional consequences of potential climatic change, what are the appropriate responses for world decision makers?

Fossil-fuel combustion is the primary source of CO_2 emissions, and also emits other "greenhouse" gases. Removing these gases after combustion is a technical challenge and imposes economic penalties. Policy options to curb emissions, so far, have stressed energy efficiency and conservation, sequestering of atmospheric CO_2, market-oriented strategies such as carbon taxes, emissions trading, switching to less CO_2-intensive fuels, and substituting renewable and nuclear energy. A warmer climate might also result in less energy consumption during winter months.

Congress has reviewed scientific information about climate change to evaluate potential economic and strategic impacts of a warmer, and perhaps more variable, climate to formulate policy responses. Because of the global implications of this problem, concerns are addressed internationally through direct communication between U.S. decision makers and world leaders, participation in international conferences, passage of legislation, and exchange of views and information with international organizations within and outside the United Nations system.

The 1992 U.N. Framework Convention on Climate Change called for a "non-binding" voluntary aim for industrialized countries to control atmospheric concentrations of greenhouse gases by stabilizing their emissions at 1990 levels by the year 2000. The 1997 U.N. Kyoto Protocol on Climate Change goes further and, if it were to enter into force, would commit world governments to legally binding emissions reduction.

MOST RECENT DEVELOPMENTS

At the November 2000 conference of parties to the U.N. Framework Convention on Climate Change (COP-6), held in The Hague, Netherlands, international climate change negotiations on implementing the Kyoto Protocol were viewed as reaching a stalemate over carbon sink issues involving mainly the U.S./Umbrella Group and the European Union. Nevertheless, the President of COP-6 concluded that major progress was made in completing many important elements of the 2-year Buenos Aires Plan of Action that would potentially govern implementation of the Kyoto Protocol. Attempts were made to bring the two major opposing blocs back to the table in Ottawa, Canada during the first week of December, although these talks did not lead to any further agreements. In any case, many FCCC parties have expressed confidence that negotiations could be resumed possibly as early as May/June 2001, at a session in Bonn, Germany, which will be called "COP-6 bis."

The FY2001 Appropriations Act for the Department of Agriculture (P.L. 106-387), enacted October 28, 2000, contained a ban on implementing the Kyoto Protocol prior to advice and consent of the Senate to ratification and would apply to certain types of climate change funding. Similar language is also found in Interior, Energy and Water, VA-HUD, Foreign Operations, and Commerce, Justice, and State appropriations bills for FY2001, some of which have become public law (see **Legislation).**

BACKGROUND AND ANALYSIS

Global Climate Change: Science and Policy

A large number of scientists believe that human activities, which have increased atmospheric concentrations of carbon dioxide (CO_2) by one-third over the past 100 years, may possibly be leading to an increase in global average temperatures. However, this so-called "global warming" theory is not without challengers who argue that scientific proof is incomplete or contradictory, and that there remain many uncertainties about the nature and direction of Earth's climate. Nevertheless, concern is growing that human activities, such as the burning of fossil fuels, industrial production, deforestation, and certain land-use practices are increasing atmospheric concentrations of carbon dioxide (CO_2) that, along with increasing concentrations of other trace gases such as chlorofluorocarbons (CFCs), methane (CH_4), nitrous oxide (N_2O), hydro fluorocarbons (HFCs), per fluorocarbons (PFCs) and sulfur hexafluoride (SF_6), may be leading to changes in the chemical composition and physical dynamics of Earth's atmosphere, including how heat/energy is distributed between the land, ocean, atmosphere and space.

Greenhouse Gases: Sources and Trends

Scientists have found that the four most important variable greenhouse gases, whose atmospheric concentrations can be influenced by human activities, are carbon dioxide (CO_2), methane (CH_4), nitrous oxide (N_2O), and chlorofluorocarbons (CFCs). Historically, CO_2 has been the most important, but over the past several decades other gases have assumed increasing significance and, collectively, are projected to contribute about as much to potential global warming over the next 60 years as CO_2. The 1997 U.N. Kyoto Protocol on Climate Change, if it were to become a treaty in force, would also regulate three other trace gases: hydro fluorocarbons (HFCs), per fluorocarbons (PFCs), and sulfur hexafluoride (SF_6), whose limited concentrations in the atmosphere are anticipated to grow over the long-term. Sulfate aerosols, a byproduct of air pollution, and other natural phenomena, are also viewed as important for their transient and regional "climate cooling" effects in Earth's atmosphere.

The amount of carbon cycling from naturally occurring processes each year through the biosphere as CO_2 is enormous – some 800 billion tons. Ice cores and other proxy climate data, which also indicate CO_2 concentrations in the atmosphere, have shown, in general, a relatively stabile global climate, at least over the past 10,000 years. As such, many scientists suggest that the amount of CO_2 generated by natural processes is about equal to the amounts absorbed and sequestered by natural processes. However, human activity since the Industrial Revolution (c.a. 1850), and primarily in the form of burning fossil fuels, is now generating some additional 24 billion tons of CO_2 per year. Available evidence shows that about half this amount is absorbed by natural processes on land and in the ocean, and that atmospheric concentrations of CO_2 are now about 32% higher than they were some 150 years ago. Some scientists suggest that a large amount of CO_2 may be stored in northern latitude soils and in temperate and tropical forests, suggesting a greater importance of the role of natural resources management and land-use practices in these regions, including burning of biomass and deforestation. Scientists estimate that anthropogenic emissions of CO_2 alone may account for as much as a 60% increase in global mean temperatures of 0.9° F, since 1850. For more information on the science of global climate change, visit the CRS Electronic Briefing Book: *Global Climate Change* website. [http://www.congress.gov/brbk/html/ebgcc1.html].

State-of-the-art computer models of the Earth's climate (GCMs) have projected a globally averaged warming of 3 to 8 degrees F over the next 100 years, if greenhouse gases continue to accumulate in the atmosphere at the current rate. Prominent climate scientists believe that such a warming could shift temperature zones, rainfall patterns, and agricultural belts and under certain scenarios, and cause sea level to rise. They further predict that global warming could have far-reaching effects – some positive, some negative depending how it may be experienced in a given region – on natural resources; ecosystems; food and fiber production; energy supply, use, and distribution; transportation; land use; water supply and control; and human health.

So-called "skeptics" of the global warming theory have called into question the reliability of the computer climate models and their output used to make projections of future warming that supported Kyoto Protocol negotiations. They also challenge some scientists' assertions that, although recent episodic weather events may seem more extreme in nature, this is indicative of long-term climate change. The Clinton Administration received criticism about attributing seemingly more frequent weather anomalies to a warming of the climate. And so the scientific questions remain: Can scientists now confirm that humans are indeed, at least in part, the cause of recent climate changes? Also, as a result of this, is the Earth committed to some degree of future global warming? If so, then what might be the consequences, and what if any of those might be prevented?

Evidence of natural variability of climate is large enough that even the record-setting warmth at the end of the 20th century does not allow a vast majority of knowledgeable scientists to state beyond a reasonable doubt that weather extremes experienced over the past two decades are attributable to "global warming," at least at the present time. However, the warming trend at the surface appear to continue. In some cases, casual relationships between seasonal and inter annual climate changes and present-day severe weather events are beginning to be recognized and even predicted, because of an improved ability to observe the *El Nino* and *La Nina* phenomena. This notwithstanding, singular extreme weather events have focused public attention on possible outcomes of potential long-term climate change and a need for a better understanding of regional climates on decadal to century scales.

National Oceanic and Atmospheric Administration's (NOAA) researchers reported that the 12 warmest years (globally averaged) since historical records have been kept occurred in the past two decades, with 1990 and 1998 among the warmest. At least some of this warming, they concluded, is human-induced. On the other hand, satellite instruments – which measure indirectly average temperatures of the atmosphere in a deep column above the surface – for the past 20 years are hard pressed to demonstrate any positive trends. A recent report by the U.S. National Research Council, Board on Atmospheric Sciences and Climate, Panel on *Reconciling Observations of Global Temperature Change* (January 13, 2000) revisited observed surface warming data of the Earth during the past 20 years. The report attempted to resolve apparent disparities between temperature data measured at the surface and those from satellites. Skeptics claim that disparate trends invalidate the output of general circulation models (GCMs), many of which demonstrate homogeneous warming throughout all the levels of the Earth's atmosphere. Panel scientists concluded that there may be a systematic disconnect between the upper and near surface atmosphere and cited physical processes which may have an unique impact on the upper atmosphere that are not currently accounted for in GCMs. In addition, they acknowledged that only that long-term, systematic monitoring of the upper atmosphere could resolve the differences in temperature trends.

A May 2000 draft report by the IPCC working group on the science of global climate change concluded "There has been a discernible human influence on the climate." However, authors reported little else new in terms of future climate change projections or the ability to resolve uncertainties about GCMs, including the behavior of clouds, or the climatic effects of human's burning of fossil fuels. Skeptics have not denied a human role in climate change, but lately have been emphasizing a modest role and an inferred minimal, if not beneficial, future impact of climate change. The third IPCC assessment was used to guide international negotiations on climate change on the Kyoto Protocol, at COP-6.

A June 2000 draft assessment report, *Climate Change Impacts on the United States: The Potential Consequences of Climate Variability and Change,* released by the U.S. Global Research Program received criticism from many of those who were involved in its early review. Critics claimed that many of the projected impacts of climate change were overstated and unsubstantiated. The National Assessment Synthesis Team (NAST), which authored the report, countered that much of the criticism it has received does not take into account the time scales upon which the report was based; the report targeted the effects of climate toward the middle of this century to the end of the next. Also, while seemingly contradictory results were produced by the two models selected for the study, NAST noted that one model scenario demonstrated a 3-4 times increase in atmospheric concentration levels of CO_2, which scientists have projected for the end of the next century, if a business as usual, or no policy change is assumed. Various regional and resource-focused assessments are not available at the USGCRP website [http://www.nacc.usgcrp.gov]. A final report by the NAST, of the same title and consisting of an overview of all of the regional and sectoral studies, was released in December 2000, although its conclusions did not differ significantly.

In August 2000, NASA scientist James Hansen suggested that climate change benefits could be achieved through near-term regulation of non-CO_2 greenhouse gases. He proposed that reducing emissions of halocarbons (refrigerants), methane, nitrogen oxides, and carbon-black aerosols (soot) could have the effect of reducing ozone (smog), in the troposphere, which itself is a greenhouse gas. Non-CO_2 greenhouse gases have relatively short atmospheric lifetimes compared with CO_2; however, most have a much larger global warming potential (gwp). This would suggest that controlling emissions of these greenhouse gases could reduce the rate and overall amount of climate warming from greenhouse gases, leaving only that expected from long-term CO_2 emissions whose full effects would not be realized for another 75-100 years hence. Nevertheless, Hansen emphasized that any actions to reduce emissions of these gases would need to be taken concomitantly with long-term strategies to reduce CO_2. Hansen also noted that modest gains from reducing CO_2 and non-CO_2 emissions in the near-term could be achieved primarily through cleaner energy production.

The Policy Context

The prospect of global warming from an increase in greenhouse gases has become a major science policy issue during the past 15 years. Seeking answers to a number of questions – How much warming?...How soon?...Should we worry? – a growing number of policymakers continue to debate the advantages and disadvantages of an active governmental role in forging policies to address prospective climate change. How real is the human-induced global warming threat? Another 10-15 years of continued warming might validate the scientific projections, but many scientists caution that waiting for this added assurance might put society at risk for a larger dose of climate change than if actions to curb or slow the buildup of greenhouse gases were implemented now. But actions on what scale?

Policymakers, here and abroad, are counseling cautious courses of action to address the prospect of climate change that many believe is still theoretical and cannot be foreseen with confidence. Given uncertainties about the timing, pace, and magnitude of global warming projections and the imprecise nature of the regional distribution of possible climate changes, and recognizing the complex feedback mechanisms within the climate system that could mask, mimic, moderate, amplify, or even reverse a greenhouse-gas-induced warming, the question is posed: What policy responses, if any, are indicated, now, or in the future?

Many proponents for early actions to address potential climate change have suggested adopting a "precautionary principle" comprised of a number of anticipatory, yet flexible policy responses that might be likened to the purchase of an insurance policy to hedge against some risks of potential climate change in the future. Broader national responses might range from engineering countermeasures, to passive adaptation, to prevention, and pursuit of an international law of the atmosphere. One policy widely advocated by President Bush in the early 1990s, and to some degree implemented to date, is the so-called "no regrets" approach, which in theory would not only reduce emissions of greenhouse gases but provide other benefits to society as well. Such policy options stress energy efficiency and conservation, increased renewable energy use, planting trees to enhance CO_2 sequestration from the atmosphere, and substitution of lesser or non-CO_2 producing fuels. Many scientists suggest that early actions might but time to gain a better understanding of global climate change and perhaps reduce possible negative impacts attributable to human-induced climate change, should they occur.

On October 19, 1993, President Clinton released his *Climate Change Action Plan* (CCAP), which proposed voluntary domestic measures to attain greenhouse gas emissions stabilization as outlined under the terms of the U.N. FCCC (see **International Action**). The CCAP reflected the President's own goals to stabilize U.S. emissions at 1990 levels by the year 2000, and called for a comprehensive suite of voluntary measures by industry, utilities and other large-scale energy users. CCAP stressed energy-efficiency upgrades through new building codes in residential and commercial sectors, and other improvements in energy generating or using technologies. Large-scale tree planting and forest reserves were encouraged to enhance sequestration of carbon dioxide and to conserve energy. Other aspects of the plan addressed mitigation of greenhouse gases other than CO_2. By avoiding mandatory command and control measures, CCAP, in one sense, appeared to be moving aggressively to implement "no-regrets" policies endorsed by former President Bush.

However, periodically, the Clinton Administration hinted at stronger regulatory actions; and some economists have suggested implementation of some form of carbon (or other energy use) tax to deter fossil fuel consumption. However, national energy taxes have historically proven to be controversial with U.S. energy producers and consumers alike. In deliberations over U.S. policy in international negotiations on global climate change, some trade groups and labor unions representing America's heavy industry, utility, and agricultural sectors have been some of the strongest vocal opponents of regulation of CO_2 emissions, claiming their members would bear the greatest economic burden of regulating fossil fuel emissions. These organizations project the loss of many American jobs to countries which would not be required to impose as strong environmental regulations, and have expressed opposition to any effort by the President to commit to greenhouse gas reductions that are not supported by sound scientific and economic analysis. Such interest groups, and some Members of Congress, have continued to challenge greenhouse gas control proposals under the 1997 U.N. Kyoto Protocol that would *not* apply to developing countries in kind, and, consequently, many of the same are opposed to U.S. ratification of the Protocol.

Not all business/industry-related organizations, the Pew Center for example, are of the same opinion, however. Some industries see an opportunity to develop and market environmental "friendly" technologies to be marketed internationally, or to switch to less CO_2-intensive fossil fuels, expand renewable and alternative energy resources for power generation, and expand use of nuclear power. Also, in efforts to garner support for or against Kyoto Protocol ratification, petitions have been circulated to thousands of scientists by major interest groups with differing views on the treaty.

Clinton Administration climate change policy encouraged voluntary efforts by government, industry and citizens alike which emphasize *flexibility* in achieving U.S. greenhouse gas emissions goals, taking into account *where* global emissions occur and *when* such reductions would be the most economically feasible. This policy addressed the life cycle and potential market of new capital equipment, e.g., energy generating technologies, that might portend savings in energy costs while enabling concomitant emissions reductions. In concert with the *when* and *where* policy, is *joint implementation* that would allow industrialized countries to share credits for emissions reduction with developing host countries. The latest dimension of the "flexible" policy response was the *what* factor, which U.S. representatives characterize as, choosing what off-the-shelf mitigation technologies, or what adaptation strategies may make the most sense to develop and utilize it now where and when feasible. While some economists have suggested that stronger climate protection measures could actually benefit the U.S. economy, by providing economic growth and employment, others such as WEFA (formerly, Wharton Economic Forecast Associates) have projected dire economic consequences, including major loss of GDP, and often conflicting results supporting both sides of the issue have depended upon what *assumptions* underlay their respective economic models.

On November 12, 1998, President Clinton instructed a representative to sign the Kyoto Protocol to "lock-in" U.S. interests achieved during negotiations. This act drew protest by some in Congress because the Kyoto Protocol had not yet, and still has not been submitted to the U.S. Senate for advice and consent to ratification. Some members claimed Clinton's action was in violation of the June 1997 Byrd/Hagel Resolution (S.Res. 98), that required an economic analysis of legally binding emission reductions on the United States, as well as participation of all FCCC parties. The President announced he would continue to pursue "meaningful" commitments from key developing countries, perhaps unilaterally, over the next few years, before he would send the treaty to the Senate in deference to S.Res. 98.

The Clinton Administration did release an economic analysis (July 1998), prepared by the Council of Economic Advisors, that concluded that with emissions trading among the Annex B-countries, and participation of key developing countries in the "Clean Development Mechanism" – which grants the latter business-as-usual emissions rates through 2012 – the costs of implementing the Kyoto Protocol could be reduced as much as 60% from many estimates. Other economic analyses, however, prepared by the Congressional Budget Office and the DOE Energy Information Administration (EIA), and others, demonstrated a potentially large loss of GDP from implementing the Protocol. Some have questioned the "hot air issue" surrounding proposed emission trading credits from joint implementation (JI) and whether these would actually be available for trade, especially in light of Eastern and Central Europe's and some countries of former Soviet Union's desire to resume rapid economic development. Furthermore, at the Ministerial session at COP-5, the EU demanded that industrialized nations' greenhouse gas emissions be reduced domestically first, in effect imposing a cap on emissions credits granted for developing country projects under JI. This continues to be a contentious topic of debate during Kyoto Protocol negotiations.

On June 3, 1999, President Clinton issued Executive Order (E.O.) No. 13123, that called for a "Greening the Government Through Efficient Energy Use." The Department of Energy has since announced that efforts under this E.O., along with other voluntary climate change initiatives undertaken to date, have helped the United States reduce its overall greenhouse emissions by as much as 19% below 1990 levels, well ahead of the timetable proposed by the Kyoto Protocol.

On November 11, 2000, President Clinton issued a statement on "Meeting the Challenge of Global Warming" in response to the results of the report: *Climate Change*

Impacts on the United States: The Potential Consequences of Climate Variability and Change (see [http://www.gcrio.org/National Assessment/]). In his statement, President Clinton said he would promulgate new regulations for U.S. electric power plants, imposing emissions caps on sulphur, nitrogen oxides, mercury, and CO_2. He also called for establishment of a domestic emissions trading program and promised a continued U.S. leadership role in climate change to set an example for other industrialized countries. Clinton announced he would take such steps as necessary to keep the United States on target for meeting Kyoto Protocol goals, if certain concessions were made regarding international adoption of flexible mechanisms such as

Emission trading, the clean development mechanism (CDM), credit for carbon sinks, and accountable, legally-binding, compliance mechanisms.

Global Climate Change Funding

On February 14, 2000, the Clinton Administration detailed plans to spend some $4.1 billion in FY2001 for climate change-related domestic programs, and investments and tax incentives. This funding included $1.7 billion for the U.S. Global change Research Program (USGCRP), which will focus on 1) Improved Climate Observations, 2) the Global Water Cycle, 3) Ecosystem Changes (climate change impacts), and 4) Understanding the Carbon cycle. USGCRP funding is divided among nine federal agencies, details of which are to be included in "Our Changing Planet: FY2001," budget document. The President recently announced a 50% increase in funding for all climate change programs from FY1998 to FY2001.

As part of this, the President slated $2.4 billion for a "Climate Change Solutions" (CCS) initiative, which includes technology investment spending and tax breaks ($4 billion in total tax incentives would be realized between FY01-FY05) associated with the President's Climate Change Technology Initiative (CCTI). Also, included under CCS were Biofuels & Bioproducts Initiatives, Energy Conservation programs, and International Clean Energy Initiative (ICEI) for international investments in "enviro-friendly" technologies, for which grants would be provided by the Agency for International Development (AID).

INTERNATIONAL ACTION

The United States was involved in negotiations and international scientific research on climate change prior to ratifying the 1992 U.N. Framework convention on Climate Change (FCCC). This included passage of a National Climate Program Act of 1978 (P.L. 95-367).

U.N. Framework Convention on Climate Change (FCCC)

The U.N. Framework Convention of Climate Change (FCCC) was opened for signature at the 1992 UNCED conference in Rio de Janeiro ("The Earth Summit"). On June 12, 1992, the United States, along with 153 other nations, signed the FCCC, that contained a legal framework that upon ratification committed signatories' governments to a voluntary "non-binding aim" to reduce atmospheric concentrations of greenhouse gases with the goal of "preventing dangerous anthropogenic interference with Earth's climate system." These actions were aimed primarily at industrialized countries, with the intention of stabilizing their emissions of greenhouse gases at 1990 levels by the year 2000; and other responsibilities would be incumbent upon all FCCC parties. On September 8, 1992, President Bush transmitted the FCCC for advice and consent of the U.S. Senate to ratification. The Foreign Relations Committee endorsed the treaty and reported it (Senate Exec. Rept. 102-55) October 1, 1992. The Senate consented to ratification on October 7, 1992, with a two-thirds majority vote. President Bush signed the instrument of ratification October 13, 1992, and deposited it with the U.N. Secretary General. According to terms of the FCCC, having received over 50 countries' instruments of ratification, it entered into force March 24, 1994.

COP-1, The Berlin Mandate

Seeking grounds for a uniform approach toward climate protection, the Conference of Parties (COP) to FCCC met for the first time in Berlin, Germany in the spring of 1995, and voiced concerns about the adequacy of countries' abilities to meet commitments under the Convention. These were expressed in a U.N. ministerial declaration known as the "Berlin Mandate," which established a 20 year Analytical and Assessment Phase (AAP), to negotiate a "comprehensive menu of actions" for countries to pick from and choose future options to address climate change which for them, individually, made the best economic and environmental sense. Criticism was leveled by many industrialized countries, including the United States, at newly industrializing countries, such as Brazil, India, and China. These would continue to be classified as non-Annex I countries and enjoy exemption from any future, legally binding emissions reduction agreements even though, collectively, these would be the world's largest emitters of greenhouse has emissions 15 years hence. (See, CRS Report 96-699, *Global Climate Change: Adequacy of Commitments Under the U.N. Framework Convention and the Berlin Mandate.*)

COP-2 and a Ministerial Declaration

The Second Conference of Parties to the FCCC (COP-2) met in July 1996 in Geneva, Switzerland. Its Ministerial Declaration was adopted July 18, 1996, and reflected a U.S. position statement presented by Timothy Wirth, former Under Secretary for Global Affairs for the U.S. State Department at that meeting, which 1) accepted outright the scientific findings on climate change-proffered by the Intergovernmental Panel on

Climate Change (IPCC) in its second assessment (1995); 2) rejected uniform "harmonized policies" in favor of flexibility; and 3) called for "legally binding mid-term targets." Legally, the Declaration represented a consensus that parties to the FCCC would not object to a "future decision which would be binding on all parties," opening the door for a possible international regulatory protocol. Individual party's objections were recorded.

COP-3, The UN (Kyoto) Protocol on Climate Change

The U.N. Kyoto Protocol on Climate Change was adopted by the COP, in December 1997 in Kyoto, Japan, one day after the official session ended. Most industrialized nations and some central European economics in transition (all defined as Annex B countries) agreed to legally binding reductions in greenhouse gas emissions of an average of 6%-8% below 1990 levels between the years 2008-2012, defined as the first emissions budget period. The United States would be required to reduce its total emissions an average of 7% below 1990 levels.

The Clinton Administration has attempted to implement emissions reductions agreed upon in the UNFCCC, and in his FY2001 budget requested funding for a Climate Change Technology Initiative (CCTI) first introduced in his FY1999 budget. Somewhat reduced funding for his climate technology initiatives has been received in previous years, and the market-based incentives, including tax breaks for industry and individuals and R&D credits for industry, still have not been implemented by Congress.

COP-4, Buenos Aires

COP-4 took place in Buenos Aires in November 1998. Here, FCCC parties adopted a 2-year "Plan for Action" to advance efforts and to devise mechanisms for implementing the Kyoto Protocol. FCCC parties also addressed compliance and financial response mechanisms to encourage more developing countries to sign on to the protocol. Talks on compliance stressed a front end "qualifying" approach rather than "sanctions and punitive measures," as the European Union (EU), and the U.S. had originally, supported. (That is, parties must be in compliance with existing commitments to take part in emission trading and joint implementation. This meant being accepted for Annex-B status and committed to terms of the Kyoto Protocol. On the other hand, few restrictions would apply for developing countries wishing to participate in the "clean development mechanism." Work continued at COP-4 to determine how to calculate emissions reductions from strengthening "carbon sinks," and devising technical definitions for sink capacity of current forest, vegetation, and land-use practices.

COP-5, Bonn, Germany

The 5th Conference of Parties to the U.N. Framework Convention on Climate Change met in Bonn, Germany, between October 25 and November 4, 1998. COP-5 included sessions of the Subsidiary Bodies on Implementation and Science and Technology and a two-day ministerial session. Major themes of negotiations included devising the technical and political mechanisms, such as the Clean Development Mechanism (CDM), Joint Implementation (JI), and developing criteria for project eligibility, all processes that would allow both developed and developing countries to meet their respective responsibilities under the FCCC, and 1997 Kyoto Protocol, with optimum flexibility. Also under consideration were legally-binding consequences for non-compliance of parties under the voluntary FCCC. This action, in and of itself, would require an amendment to the Kyoto Protocol, as well as establishment of a COP-certified national inventorying systems to track international greenhouse gas emissions and their reduction. Parties adopted a decision for a second round of national communications and emissions reporting (for Annex I countries); so that updated data and information could be used to inform upcoming negotiations at COP-6.

COP-6, The Hague, Netherlands

The Sixth Conference of Parties to the United Nations Framework Convention on Climate Change (COP-6) convened November 13-25, 2000. Despite a major impasse reached at this session for final implementation of the Kyoto Protocol, a number of FCCC parties expressed confidence that progress was made in resolving a number of technical issues associated with the Kyoto Protocol. Throughout the session, however, the United States and European Union (EU) parties remained split along policy lines as to the content of the final treaty and how it should be implemented. A proposal submitted by the United States focusing on forestry and land-use practices and land-use change as a possible means to receive legitimate emissions reduction credits was also considered; however, who might take those credits and when they might be applied was contested among FCCC parties. Talks continued in Ottawa, Canada during the first week of December 2000, but no further agreements were reached.

Throughout COP-6 many parties seemed in favor of developing actual mechanisms by which Annex B countries under the Kyoto Protocol might trade emissions credits and pursue flexible response strategies such as international joint implementation among industrialized (Annex 1) countries; and also develop a clean development mechanisms (CDM) in cooperation with industrialized countries, for lesser developed countries that might allow them to forego greenhouse gas emissions-intense economic development paths.

However, negotiations appeared to falter when the EU charged that the United States stood to enjoy a "number of loopholes" under the agreement negotiated thus far and would lessen the former's respective burden of domestic emissions reductions. Instead, the EU sought to impose certain limits on the United States in using these flexibility mechanisms. Otherwise, the EU claimed that little would be accomplished in terms of environmental improvement if the United States failed to rein-in its contributions of greenhouse gas emissions (24% of the total). U.S. negotiators countered that U.S. emissions growth had been stemmed significantly over the past 10 years, growth was occurring elsewhere, and that emissions reductions should be counted no matter where in the world they occur or how they are achieved. Also under debate was whether some U.S. proposals would achieve any real emissions reductions. For example, the EU consistently criticized the United States for its proposals to pay Russia and Central European countries for rights for it to pollute, if surplus credits were eventually granted to the latter, under Annex B of the Kyoto Protocol

Already, a number of industrialized countries, including the U.S. and the Umbrella Group consisting of Japan, U.S., Canada, Australia, New Zealand, Russia, the Ukraine and Norway, and their major industries, are proposing domestic schemes for emissions trading, which might at some point, negotiators claim, serve as a model for an international trading regime under an international agreement sanctioned by UNFCCC parties. The EU even has its own proposal for such a scheme, know as the "EU Bubble."

In the opinion of many business leaders and some Members of Congress, current domestic proposals for voluntary flexible mechanisms to reduce global greenhouse emissions have had strong U.S. business and agricultural community input and support; this support has been reflected in recent legislation proposed and supported in a bi-partisan spirit in Congress. (See **Legislation**). On the other hand, there was strong opposition to the Kyoto Protocol by the U.S. congressional delegation, which attended COP-6. Regardless of the ultimate legal vehicle – the Kyoto Protocol, or some other treaty – in order for the United States to move forward on any such agreements will require the advice and consent of the U.S. Senate before any such agreements are binding for U.S. citizens. Congress would also have the responsibility to implement provisions of the treaty, domestically, if in force.

Major environmental interests continue to oppose any agreement that would not take strong international legally-binding regulatory action to reduce greenhouse gas emissions. On the other hand, many reputable scientists feel that the scientific evidence about climate change has been compelling enough for world leaders to accept prospective climate change as a certainty. To reduce the apparent human contribution, they believe, can only be done by reducing atmospheric emissions of greenhouse gases under an international regulatory regime regardless whether those remedial actions would be voluntary or mandatory. In September 2000, world leaders adopted the text of the "U.N. Millennium Declaration," which affirmed nations should make efforts to ensure that the Kyoto Protocol enters into force by the next U.N. Conference on Economic Development (UNCED), scheduled for 2002. At the same time skeptics of global warming continue to question whether any action is necessary based upon their interpretation of the scientific data.

Most importantly, the Kyoto Protocol has yet to be ratified by the United States and is still a long way from meeting the criteria established by the COP necessary for its entry into force, and there are a number of policy issues confronting approval of the Kyoto Protocol that still need to be resolved. Some of these include: 1) deciding what parties will be able to participate in flexibility mechanisms should they be adopted (and who *should* participate); 2) determining the adequacy of current commitments of parties under the FCCC; 3) establishing what might be the consequences, if any, for non-compliance under a prospective protocol; 4) determining whether the amount of emissions credits earned by any one country should be capped, if resulting in nothing more that a zero-sum gain for global emissions reductions; 5) where and when carbon sink credits might be taken; and 6) whether activities primarily taken in a host country should be credited in the absence of any domestic efforts (the so-called "additionality" argument) – and the crux of the EU's resistance to U.S. flexibility mechanisms.

Many developing countries lacking the necessary monetary and human resources fear that they will face significant impacts from climate change. Many of these countries have expressed frustration over the results of COP-6 because of what they now believe are negotiations which are not in their interest, but being held hostage by the interests of the United States and EU.

CONGRESSIONAL INTEREST AND ACTIVITIES

No fewer than 17 bills were introduced on global climate change in the first session of the 106th Congress, but only a handful received consideration in the second session. Many of these dealt with congressional concern about possible "back door" implementation of the 1997 Kyoto Protocol to the 1992 U.N. Framework Convention on Climate Change, that is, implementation prior to advice and consent to ratification by the Senate. Related directives were found in other FY2001 appropriation bills. Other climate change legislation dealt with funding priorities for U.S. research activities and still other proposed schemes for voluntary greenhouse gas emissions reductions and greenhouse gas emission trading credits. Also President Clinton requested increases for a number of presidential climate change initiatives for FY2001. (See **Global Climate Change Funding.**)

Most of the bills not related to appropriations introduced this session seek ways to act legislatively on climate change protection, in lieu of the Kyoto Protocol, and these mostly stress voluntary measures. Much legislative activity in the 106th Congress was confined to relevant committees of jurisdiction, or had been in the nature of oversight hearings, Also, Congress continued to exercise its oversight authority to ensure that aspects of U.S. research on climate change are complying with the 1994 Government Performance and Results Act (GPRA), and that such research is being used effectively to inform the policy process. The Senate still awaits transmittal of the Kyoto Protocol for its advice and consent ratification.

New scientific findings concerning the human contribution to climate change emerged during expert review of the third IPCC assessment on climate change, recently the overall projections of temperature and sea-level rise made in the 1995 IPCC Assessment were estimated to be higher than previously reported, and mostly on the high end of predictions. The IPCC has also suggested that it may be prudent to consider other potential greenhouse gases not slated to be regulated by the Kyoto Protocol, and also to account for potential indirect climatic change effects that may be attributable to other atmospheric emissions (e.g., replacements for ozone depleting substances).

On January 13, 2000, the National Research Council released a report which attempted to reconcile different surface and atmospheric temperature trends and the implication for global climate change models (GCMs), and confirmed a positive temperature trend at the surface since 1970. On November 10th, the USGCRP released its assessment of the potential consequences of climate change impacts on the United States; the results, the National Assessment Synthesis team called for action to address potential significant regional climate changes in the United States resulting from global climate change. These two reports, and many other issues stated above, will likely be subject to House and Senate hearings in the 107th Congress.

LEGISLATION

P.L. 106-387, the Agricultural Appropriations Act for FY2001 (H.R. 4461), and its conference report (H.Rept. 106-619 § 734) contain language about funding for Kyoto Protocol activities. Some have argued its provisions challenge programs that were previously authorized by Congress in other laws. H.R. 4461 served as a model for language in other climate change appropriations bills and reports. The majority of these prohibit unauthorized finding of activities related to the Kyoto Protocol prior to the Senate's advice and consent to ratification of this Treaty. In many cases, technical and clarifying amendments were added to these bills as they have progressed through the FY2001 appropriations process to ensure climate change programs previously authorized by Congress continue to receive funding.

S. 882 (Murkowski) Strengthens provisions in the Energy Policy Act of 1992 and the Federal Non-nuclear Energy Research and Development Act of 1974 with respect to potential Climate Change. Established the Office of Global Climate Change in Department of Energy for coordination of climate change activities at agency which reports to Secretary of Energy. Authorizes $2 billion for a ten-year research, development and demonstration program to develop new technology through public-private partnerships to help stabilize greenhouse gas concentrations in the atmosphere. Promotes voluntary efforts to reduce greenhouse gas emissions and reporting under Sect. 1605 of the Energy Policy Act of 1992. Reiterates continued opposition to Kyoto Protocol because of its potential costs and little, if any environmental gain. Introduced April 27, 1999; referred to Committee on Energy and Natural Resources. Hearings held March 30, 2000.

S. 935 (Lugar) National Sustainable Fuels and Chemicals Act of 1999. Amends the National Agricultural Research, Extension, and Teaching Policy Act of 1997 to authorize research to promote conversion of biomass into bio-based industrial products, and for other purposes. Cites zero net greenhouse gas emissions from biomass fuels. Authorizes $49 million per year for each fiscal years 2000 through 2005 for a sustainable fuels and chemicals research initiative, including research on bio-based products that can compete in performance with fossil-based products, and on accurate measurement and analysis of carbon sequestration and carbon-cycling in relation to bio-based industrial products and feed-stocks. Introduced April 30, 1999; referred to Committee on Agriculture, Nutrition, and Forestry. Hearings held May 27, 1999. Ordered to be reported with amendments in the nature of a substitute on July 29, 1999; Senate Agriculture Committee reported measure (S.Rept. 106-179) on October 8, 1999. S. 935 passed the Senate (amended) on February 29, 2000, with an amendment to the Title, and was referred to the House Agriculture and Science Committees on March 9, 2000. Referred to House Subcommittee on Energy and Environment on March 23, 2000.

S. 1066 (Roberts) Introduced May 18, 1999, referred to Senate Committee on Agriculture, Nutrition and Forestry. Amends the National Agricultural Research, Extension, and Teaching Policy Act of 1977. Encourages the use of and research into agricultural best practices to improve the environment, and for other purposes. Findings of bill cite magnitude and importance of carbon storage in soils. Authorizes appropriations of $10 million. Suggests enhancement of carbon storing strategies through agricultural best practices in lieu of implementing the Kyoto Protocol. Requires the Economic Research Service to report on the impact on the farm economy of the United States under the Kyoto Protocol and Framework Convention on Climate Change. Allows for research MOU between USDA and NOAA. On June 20, 2000, the House ordered the bill reported with an amendment in the nature of a substitute. On September 2, 2000, the Senate ordered the measure reported as an amendment in the nature of a substitute (S.Rept. 106-407). The House prepared this legislation for the Floor on October 18, 2000.

S. 1776/S. 1777 (Craig) Climate Energy Policy Response Act. Amends the Energy Policy Act of 1992 to revise the energy policies of the United States in order to reduce greenhouse gas emissions (voluntarily), advance global climate science, promote technology development, and increase citizen awareness, and for other purposes (e.g., creates mechanisms and institutions necessary for implementing provisions of Bill). Section 1604 extends authority to DOC to become the lead agency on climate change research and public information programs in support of these provisions. Section 1612 establishes a Natural Resource Center on Climate Change (NRCCC). Makes any certifications for emissions reductions prior to enactment subject o review by Secretary of Energy. Authorizes funding for energy technology RD&D. S.1777 (Craig) is tax legislation which would enable implementation of S. 1776. Both introduced October 25, 1999; S. 1776 was referred to Senate Committee on Energy and Natural Resources; hearing held March 30, 2000. S. 1777 was referred to Senate Finance Committee on October 25, 1999.

S. 2540/S. 2982 (Brownback) Amends the Food Security Act of 1985 to require the Secretary of Agriculture, acting through the Chief of the Natural Resources Conservation

Service, to establish a carbon sequestration program to permit owners and operators of land to enroll land in the program to increase the sequestration of carbon, educational outreach through the Agricultural Extension Service and for other purposes. Authorization to be make from funds available for climate change initiatives or greenhouse gas emission reductions. Domestic Carbon Storage Incentive Act of 2000. Introduced May 10, 2000; referred to Committee on Agriculture, Nutrition, and Forestry. Similar in intent, S. 2982 would enhance *international* conservation to promote the role of carbon sequestration as a means of slowing the buildup of greenhouse gases in the atmosphere, and promote voluntary, proactive efforts on the issue of climate change. Introduced July 27, 2000; referred to the Senate Committee on Finance.

CHRONOLOGY

11/13/00-11/25/00 – COP-6 convened in The Hague, Netherlands. USGCRP released National Assessment of Potential Impacts of the United States of Climate Change. President Clinton promised U.S. leadership for facing the challenge of climate change, and proposed new rules for pollution from U.S. electricity production and a domestic emissions trading scheme.

9/08/00-09/15/00 – Preparatory session for COP-6 in Lyons, France. U.N. Millennium Declaration proposed entry into force of Kyoto Protocol by UNCED 2002.

06/02-12/00 – UNFCCC Subsidiary Bodies met in Bonn, Germany in preparation of COP-6.

10/25/99-11/04/99 – COP-5 convened in Bonn, Germany

11/12/98 – President Clinton instructed a U.S. representative to sign the Kyoto Protocol.

11/2-13/98 – COP-4 met in Buenos Aires; a 2-year "Plan of Action" was adopted.

03/16/98 – Kyoto Protocol opened for signature.

12/01-11/97 – U.N. Kyoto Protocol on Climate Change adopted at conclusion of COP-3.

06/12/97 – S.Res. 98 introduced (the Byrd/Hagel Resolution).

7/08-19/96 – Ministerial Declaration issued in Geneva, Switzerland at COP-2 (07/18/96).

03/28-04/06/95 – First Conference of Parties (COP-1) met, adopted "Berlin Mandate."

03/24/94 – U.N. Framework Convention on Climate Change entered into force.

FOR ADDITIONAL READING

U.S. Department of State. Bureau of Oceans and International Environmental and Scientific Affairs. Office of Global Change. *Climate Action Report: 1997 Submission of the United States of America Under the U.N. Framework Convention on Climate Change.* (DOS pub. 10496), Washington, DC: 1996.

U.S. Executive Office of the President. *Climate Change Action Plan,* by President William J. Clinton and Vice President Albert T. Gore, Jr. Washington, DC: October 1993.

U.S. Global Change Research Program. *Climate Change Impacts on the United States: The Potential Consequences of Climate Variability and Change*, Overview by the National Assessment Synthesis Team. Washington, DC: November 2000.

U.S. Office of Science and Technology Policy. Committee on Environment and Natural Resources Research. *Our changing planet: the FY20000 U.S. Global Change Research Program; Implementation and Budget Overview*. Washington, DC: May 1999.

U.S. National Research Council. Commission on Geosciences, Environment, and Resources. Board of Atmospheric Sciences and Climate. Climate Research Committee. *Reconciling Observations of Global Temperature Change*, by the Panel. Washington, DC: January 2000.

Chapter 2

GLOBAL CLIMATE CHANGE: THE ROLE FOR ENERGY EFFICIENCY

Fred Sissine

INTRODUCTION

Energy efficiency is increased when an energy conversion device, such as a household appliance, automobile engine, or steam turbine, undergoes a technical change that enables it to provide the same service (lighting, heating, motor drive) while using less energy.[1] Energy efficiency is often viewed as a resource option like coal, oil or natural gas. It provides additional economic value by preserving the resource base and reducing pollution.

Energy security, a major driver of federal energy efficiency programs in the past, is now somewhat less of an issue. On the other hand, worldwide emphasis on environmental problems of air and water pollution and global climate change have emerged as important drivers of support for energy efficiency policies and programs. Also, energy efficiency is seen as a technology strategy to improve the competitiveness of U.S.-made appliances, cars, and other energy-using equipment in world markets. The Clinton Administration views energy efficiency as the flagship of its energy policy for global climate change and other environmental reasons.

From 1975 through 1985, high-energy prices served as a strong catalyst to improved energy efficiency.[2] However, the sharp drop in oil and other energy prices that began in 1986 has dampened the impact of prices on energy efficiency improvements.

Federal policies and programs have also made a significant contribution to improved energy efficiency.[3] One such program is DOE's energy efficiency R&D program, which

---

[1] A more detailed definition of energy efficiency is available in CRS Issue Brief IB10020, *Energy Efficiency: Budget, Climate Change, and Electricity Restructuring Issues.*

[2] DOE. *Energy Conservation Trends: Understanding the Factors Affecting Energy Conservation Gains and Their Implications for Policy Development.* 1995. p. 2-3.

employs a "technology-push" strategy. That is, it produces new, ever-more efficient technologies that form a basis for new products and services in the private sector. In contrast, EPA's energy star programs employ a "market-pull" strategy wherein businesses, institutions, and consumers are encouraged to buy more energy-efficient equipment.

The role of energy prices and the environmental benefits of energy efficiency often lead to a discussion about barriers and market failures. However, the resultant debate over the effectiveness of market forces to stimulate energy efficiency and the merit of federal policies and programs that support energy efficiency is not the focus of this report.[4] Instead, this paper is focused on the projected contribution of energy efficiency to reducing CO_2 emissions.

Energy efficiency is proposed as a cost-effective and reliable means for reducing the nation's growth in CO_2 emissions due to fossil fuel use. Recognition of that potential has led to high expectations for the control of future CO_2 emissions through even more energy efficiency improvements than have occurred through past programs, regulation, and price effects. Thus, in a recent context of low energy prices and rising fossil fuel use, the Clinton Administration has proposed increased government support for energy efficiency programs as its primary initiative to reduce emissions of CO_2 and other "greenhouse gases" that may cause global climate change.

However, there is a debate over projected estimates of the future potential for energy efficiency to curb the growth of CO_2 emissions through 2010. This paper discusses this debate, which is centered on differences between key reports by the Department of Energy (DOE) and the Energy Information Administration (EIA). A DOE report by five of its research laboratories projects that further gains in energy efficiency could be the largest future contributor to CO_2 emissions reduction. However, EIA has criticized the DOE report's assumptions about the character of future energy efficiency measures, economic growth rates, future government R&D policies, and market adoption of energy efficiency measures.

The paper also describes a debate over the analysis of actual CO_2 emission reductions from past energy efficiency measures. In this case, methodological issues are at the core of disagreements between the General Accounting Office (GAO) and the Environmental Protection Agency (EPA) about the best way to assess emission savings from EPA's various energy efficiency programs.

Finally, the paper notes that federal efforts to curb global climate change through increased energy efficiency may be affected by a number of issues being debated by Congress, including program appropriations, new tax incentives, and legislation on electricity restructuring.

---

[3] Ibid.
[4] This topic is discussed in CRS Issue Brief IB10020, *Energy Efficiency: Budget, Climate Change, and Electricity Restructuring Issues.*

ENERGY USE IMPACT ON GLOBAL CLIMATE CHANGE

Whenever energy efficiency and conservation measures reduce fossil fuel use, they will reduce carbon dioxide (CO_2) emissions, as well as pollutants that contribute to water pollution, acid rain, and urban smog. Human activities, particularly burning of fossil fuels, have increased atmospheric CO_2 and other trace gases.[5] If these gases continue to accumulate in the atmosphere at current rates, many experts believe global warming could occur through intensification of the natural "greenhouse effect," that otherwise moderates Earth's climate. Excess CO_2 is the major contributor to this effect. The influence of human-induced emissions on the "greenhouse effect" is a subject of continuing research and controversy.[6]

U.S. use of fossil energy (coal, oil, natural gas) currently produces about one-fourth of the world's CO_2 emissions. Since, 1988, the federal government has accelerated programs that study the science of global climate change and created programs aimed at mitigating fossil fuel-generated carbon dioxide (CO_2) and other human-generated emissions. The federal government has funded programs for energy efficiency as a CO_2 mitigation measure at DOE, EPA, the Agency for International Development (AID), and the World Bank. The latter two agencies have received funding for energy efficiency-related climate actions through foreign operations appropriations bills.[7]

Efforts to study greenhouse gas emissions and to devise programs to reduce them accelerated after the 1992 United Nations Conference on Environment and Development (UNCED) concluded with the signing of the Rio Declaration, Agenda 21 (an action program), and the Framework Convention on Climate Change (UNFCCC). Agenda 21 promotes the development, transfer, and use of improved energy-efficient technologies; the application of economic and regulatory means that account for environmental and other social costs, and other energy efficiency-related measures. The United States ratified the UNFCCC in 1992, and the Convention entered into force in 1994. The UNFCCC calls for each nation to develop a strategy for emissions reduction, inventory emissions, and promotion of energy and other technologies that reduce emissions.

ENERGY EFFICIENCY AND ENERGY USE

Increased energy efficiency of combustion and other fuel-using equipment has a long record of reducing the rate of growth in fossil fuel use and, thereby, reducing carbon emissions. This improvement is reflected in the ratio of U.S. energy use to Gross Domestic Product (GDP), which fell from 19,750 British thermal units (BTU's) per dollar in 1971 to 14,040 BTU's per dollar in 1986.[8] This represents an average annual

---

[5] The trace gases include chlorofluorocarbons (CFCs), methane, and nitrous oxide.

[6] For more on the science of climate change, see CRS Issue Brief 89005, *Global Climate Change*, by Wayne Morrissey and John Justus.

[7] For more on the foreign operations spending, see CRS Report 97-1015F. *Global Climate Change: The Role of U.S. Foreign Assistance,* by Curt Tarnoff.

[8] U.S. DOE. EIA. *Annual Energy Review 1998.* July 1999. p. 12-13. Values are expressed in 1992 constant dollars.

reduction of 1.81% in the energy/GDP ratio. For the period from 1972 to 1986, energy efficiency improvements cut energy use by 30% or 32 quadrillion BTU's per year.[9] By 1988, recognition of this accomplishment had led to a focus on energy efficiency programs as a key strategy for future control of CO_2 emissions.

However, from 1986 to 1998, the rate of energy efficiency improvement slowed. The energy/GDP ratio declined from 14,040 BTU's per dollar in 1986 to 12,480 BTU's per dollar in 1998, but this represents an average annual reduction of 0.85%, which is less than half the rate for 1972 to 1986. Further, the decline in oil prices since the mid-1980s has led to historically low gasoline prices which, in turn, encouraged motorists to buy less fuel-efficient automobiles, such as sport utility vehicles, and to increase travel by about 24%.[10] Overall, national petroleum use for transportation grew 21%, or 4.3 Q during this period. Also, since 1994, electric utility industry restructuring at the state level caused utility spending for energy efficiency to fall 48% by 1998 and the resultant rate of energy savings fell 20% from 1996 to 1998.[11] Meanwhile, coal use for electricity production grew 33% from 1986 to 1998.[12]

Thus, despite the increase in efficiency as measured by BTU/$, total fossil fuel use, has been rising steadily due to low energy prices, economic growth, and population growth. This growth includes oil and coal, which are the most intense emitters of carbon dioxide (CO_2). As a result, CO_2 emissions have been rising, eclipsing the 1993 Clinton Administration Climate Change Action Plan (CCAP) goal of reducing emissions to the 1990 level by 2000. In fact, Energy Information Administration (EIA) projections show fossil energy use and emissions increases continuing through 2010.[13]

CARBON EMISSIONS REDUCTION AND ENERGY EFFICIENCY

Climate Change Action Plan (CCAP)

In 1993, the Clinton Administration launched a Climate Change Action Plan (CCAP) that sought to stabilize year 2000 CO_2 emissions at the 1990 level of 1,346 million metric tons of carbon (MMTC). To achieve this goal, the plan relied primarily on voluntary measures for increasing energy efficiency. A variety of CCAP programs were funded at DOE, EPA, and other agencies, but at levels well below budget requests. The nation clearly did not reach the year 2000 stabilization goal. Instead, emissions rose to 1,485 MMTC in 1998 and are projected to reach 1,552 MMTC in 2000, which would be a 206 MMTC, or 15% increases above the 1990 level

---

[9] U.S. DOE. *Energy Conservation Trends.* 1989. p. 5. [DOE/PE-0092]

[10] EIA. *Monthly Energy Review.* December 1999. Table1.10. p. 17.

[11] EIA. *Electric Power Annual 1998 Volume II.* December 1999. Table 44. p. 75.

[12] EIA. *Monthly Energy Review,* Table 6.2, p. 88.

[13] DOE. EIA. Annual Energy Outlook (AEO) 2000. Reference Case Forecast. Table A19. p. 141.

Kyoto Protocol's Target for 2010

By 1995, growing worldwide recognition of the difficulty in reaching year 2000 stabilization led to meetings of the conference of parties to the UNFCCC to set enforceable targets for emission reductions for the post-2000 period. This effort culminated in December 1997, where the third conference of parties (COP-3) met in Kyoto, Japan, to set years 2008-2012 targets for emission reductions.

The 1997 Kyoto Protocol calls for: (1) the United States to reduce by 7% from baseline years (1990 for CO_2) the average annual tons of carbon equivalent released by six greenhouse gases during the period 2008-2012; (2) implementation through market mechanisms such as international joint implementation and emissions trading schemes; and (3) encouragement of "clean energy" development in developing countries. However, critics maintain that the Protocol does not require developing nations to "meaningfully participate" in the emission reduction effort. This is a major barrier to Administration goals and Senate ratification and that is fostering additional negotiations before the Administration will seek Senate ratification of the Protocol.[14]

Inventories of CO_2 emissions are fairly well established and account for about 85% of the total carbon-equivalent emissions from all six greenhouse gases. In addition to using energy efficiency and other means to curb CO_2 emissions from energy production, CO_2 can be sequestered through re-forestation and other carbon "sinks." Due to the way sinks are counted, and due to other provisions in the Protocol, the actual reduction of U.S. greenhouse emissions required to meet the Kyoto target may be less than 7% below the 1990 CO_2 baseline. The uncertainties about sinks, and larger uncertainties about future economic growth rates and other variables, create a broad range of uncertainty about the projected average level of emissions over the 2008-2012 period. For example, two major studies yielded a range from about 390 MMTC to 660 MMTC as the projected emission reduction requirement needed to achieve the U.S. goal. This represents a 70% range of uncertainty in the emissions reduction task. Given the previous failure of the 1993 CCAP to stabilize emissions by 2000, the current Kyoto target for actually reducing emissions in the 2008-2012 time frame looms as a major policy challenge, assuming Senate ratification of the Kyoto Protocol.[15]

The Kyoto Agreement set 1990 as the baseline year for CO_2 emissions, from which progress toward targets for future reductions are to be measured. DOE's Energy Information Administration (EIA) is the recognized authority for assessing actual levels, and projecting baseline "business-as-usual" (BAU) future levels, for CO_2 and all other U.S. greenhouse gas emissions. EIA has established the 1990 CO_2 level at 1,346 MMTC. EIA's Annual Energy Outlook (AEO) projects future CO_2 levels. Assuming no major future policy actions, the BAU scenario in the AEO projects a large growth in CO_2 emissions by 2010. However, in accounting for recent policy changes and projected

---

[14] For more details about the Kyoto Protocol, see the CRS electronic briefing book on Global Climate Change at [http://www.congress.gov/brbk/html/ebcctop.html].

[15] For more on the other greenhouse gases and Kyoto reduction targets, see CRS Report 98-235, *Global Climate Change: Reducing Greenhouse Gases – How Much and From What Baseline?*, by Larry Parker and John Blodgett.

economic trends, the AEO's projections vary considerably from year-to-year. The projected BAU level for CO_2 in 2010 stood at 1,730 MMTC in the 1997 AEO, 1,803 MMTC in the 1998 AEO, 1,791 MMTC in the 1999 AEO, and 1,787 in the 2000 AEO. Thus, for example, the 1997 projection for 2010 would be a 384 MMTC increase over the 1990 level. The 2000 projection for 2010 would be a 441 MMTC, or 33% increase over the 1990 level.

Energy Efficiency Impacts Projected for 2010

Because CO_2 contributes the largest share of greenhouse gas emission impact, it has been the focus of studies of the potential for reducing emissions through energy efficiency and other means. In preparation for the meeting of the Third Conference of Parties (COP-3) to the UNFCCC held in Kyoto, DOE's Office of Energy Efficiency and Renewable Energy (EERE) issued a September 1997 report by five national laboratories entitled *Scenarios of U.S. Carbon Reductions: Potential Impacts of Energy Technologies by 2010 and Beyond.*[16] Also, known as the *Five-Lab Study*, it assumed the 1990 baseline of 1,346 MMTC and used the 1997 BAU projection that emissions would reach 1,730 MMTC in 2010 – an increase of about 384 MMTC, or 29%. This is shown in Table 1. The report analyzes some options for using cost-effective high-efficiency (HE) energy technologies and other low-carbon (LC) technology options to curb emissions. It projects that a combination of HE/LC technology and a permit price of $50 per ton of carbon could bring 2010 emissions to a level just below the 1990 stabilization level. Lower permit price assumptions yielded 2010 emission levels between the BAU and stabilization levels.

The *Five-Lab Study* anticipates, as Table 2 shows, that energy efficiency is the single largest contributor to meeting U.S. CO_2 targets, accounting for 50% to 90% of the projected emissions reduction in 2010. The transportation sector yields the most reduction; from automobile weight reduction, fuel cell breakthroughs, and other options. The buildings sector yields reductions from lighting, space conditioning, and other options. The industry sector yields savings from combined heat and power (CHP), motor system design, and a variety of technologies specific to each industry, such as impulse drying for pulp and paper plants and direct smelting for iron and steel plants. In the utility sector, some savings come from improved power plant efficiency, but the largest contribution is from carbon permit prices stimulating the use of low carbon fuels. Further, the *Five-Lab Study* projects that all emission-reduction scenarios can be achieved at low or no net direct cost to the economy.

In a 1998 report, Impacts of the *Kyoto Protocol on U.S. Energy Markets and Economic Activity*, EIA finds problems with several key assumptions in the *Five-Lab Study* about the use of new energy-efficient technologies to reduce emissions. These assumptions include "…increased performance and lower costs for new technologies, new [unspecified] government policies that promote adoption into the market, and a

---

[16] Available at [http://www.ornl.gov/ORNL/Energy_Eff/labweb.htm].

greater propensity by consumers to buy them than they have shown in the past." Specific examples include use of a 15-year payback for buildings technology when consumers expect one-year to five-year paybacks; use of a 6% industrial market penetration factor in an EIA model that normally assumes 3%; and use of 50 mpg for new car fuel economy while EIA estimates about 33 mpg.

EIA further criticizes the *Five-Lab Study* for assuming an aggressive R&D program and a 1.9% annual economic growth rate, which is 10% lower than EIA's assumption of a 2.2% rate. Moreover, EIA says the *Five-Lab Study* uses a series of independent, non-integrated, end-use models that fail to capture feedback between energy markets and the rest of the economy and likely includes some double counting of emission reduction benefits. Additionally, EIA notes that none of the scenarios in the Five-Lab Study yields emissions below the 1990 level, because they were designed to achieve stabilization at 1990 levels. In contrast to the *Five-Lab Study*, EIA's equivalent scenario (see Table 1, scenario for "1990+9%") finds that a higher carbon price of $163 per MMTC would be required and that the Gross Domestic Product (GDP) would be about 2%, or $235 billion (in year 2000 constant dollars), lower.

Climate Change Technology Initiative (CCTI)

As it became clear that the CCAP would fall short of its goal of stabilizing emissions by 2000 and as the Kyoto Protocol set an updated round of goals to reduce emissions further by 2008-2012, the Clinton Administration responded by issuing its Climate Change Technology Initiative (CCTI) proposals for increased energy efficiency research and development (R&D) spending, tax credits, and other policy mechanisms at the Department of Energy, Environmental Protection Agency, and other agencies.[17,18] EPA and DOE have stressed the urgency of action, contending that CCTI provisions would provide immediate savings in energy, costs, and emissions. In contrast, DOE's Energy Information Administration has contended that the CCTI provisions would provide a minimal impact on emissions. Congress has approved only small amounts of the CCTI requests and has expressed concerns about approving the Kyoto Protocol, which would set a national target for emission reductions through 2012.

MEASURING ENERGY EFFICIENCY IMPACTS

Important issues relate to measuring or otherwise verifying a reduction of emissions from past to projected future levels.

Studies show that energy efficiency measures have slowed fossil energy demand and provided real reductions in CO_2 emissions compared to projected growth rates. There is a

---

[17] For more on the R&D proposals, see CRS Report 98-408 STM. *Global Climate Change: Research and Development Provisions in the President's Climate Change Technology Initiative,* by Michael Simpson.

[18] For more on the tax proposals, see CRS Report 98-193 E. *Global Climate Change: The Energy Tax Incentives in the President's FY2000 Budget,* by Salvatore Lazzari.

growing professional literature on the assessment of energy efficiency program impacts, which forms a basis for assessing their effect on emissions.[19] One key study is DOE's report Energy Conservation Trends,[20] which presents the most complete analysis available on the achievements of DOE energy efficiency policies. EIA has also begun to examine the analytical bases for verifying the achievements of energy efficiency.[21]

However, many of these same studies show that long-range energy savings result from a diverse array of measures whose savings are not easily disentangled from the impacts of energy prices, consumer behavior, and other variables. As a result, claims to achieving a certain amount of saving may be subjected to dispute. For example, GAO and EPA disagree about the methods used (and the resulting savings estimates) for assessing emission reductions due to EPA's CCAP energy efficiency programs.[22]

According to EPA, the Administration evaluates the effectiveness of its climate programs through an interagency program review. The first such interagency evaluation, chaired by the White House Council on Environmental Quality, examined the emissions impact of CCAP. The results were published in the U.S. Climate Action Report 1997, as part of the U.S. submission to the UNFCCC.[23] The GAO reviewed estimates of the emission-reduction impacts for four of 20 EPA voluntary programs under CCAP.[24] For two of the four programs, GAO found that EPA did not adjust emission reduction estimates to account for non-program factors that may have contributed to the reported results.[25] This critique has led to an ongoing debate between GAO and EPA over methods of measuring program impacts and the reliability and validity of reported emission reduction estimates.[26,27]

There are a number of energy efficiency measurement issues. First, is which indicators should be used to assess progress in energy efficiency? Energy use per unit of gross domestic product (GDP) is one popular measure. However, energy use per person is another very informative measure. Also, there are cause and effect questions that are difficult to assess and could be masked by the very general energy/GDP ratio and energy/person ratio. For example, are improvements to such ratios due directly to energy

---

[19] International Energy Program Evaluation Conference. *Evaluation in Transition: Working in a Competitive Energy Industry Environment.* Proceedings. 1999. 986 p.

[20] DOE. *Energy Conservation Trends: Understanding the Factors Affecting Energy Conservation Gains and Their Implications for Policy Development.* 1995. 50 p.

[21] Energy Information Administration. *Measuring Energy Efficiency in the United States' Economy: A Beginning.* (DOE/EIA-0555[95]/2) October 1995. 91 p.

[22] U.S. Environmental Protection Agency. *Energy Star and Related Programs 1997 Annual Report.* March 1998. (430-R-98-002) 37 p.

[23] U.S. Department of State. Office of Global Change. *Climate Action Report.* 1997. p. 79-82.

[24] U.S. GAO. *Global Warming: Information on the Results of Four of EPA's Voluntary Climate Programs.* 1997. p. 26. [GAO/RCED-97-163] GAO notes that the four programs – Green Lights, Source Reduction and Recycling, Coalbed Methane Outreach, and State and Local Outreach – represented about one-third of EPA's CCAP funding and about one-third of the estimated emission reductions for year 2000.

[25] Ibid, p. 2. The two programs are Green Lights and State and Local Programs.

[26] U.S. Congress. Senate. Committee on Energy and Natural Resources. Hearing. *GAO's Review of the Administration's Climate Change Proposal.* June 4, 1998.

[27] U.S. EPA. Climate Protection Division. *Driving Investment in Energy Efficiency: Energy Star and Other Voluntary Programs.* [EPA 430-R-99-005] July 1999. 35 p. Reports on EPA's latest estimates of emission reductions from its energy efficiency programs.

efficiency R&D, energy efficiency programs and policies, energy prices, productivity enhancements, or consumer behavior? Also, there are a variety of methods that can be applied to seek answers to these questions. They include simulation models, economic models, program impact evaluations, and others.

An effort is underway to create an international standard for measuring savings from energy efficiency. DOE and other agencies collaborated to create the International Performance Measurement and Verification Protocol (IPMVP).[28] Its purpose is to provide a common technical language for assessing the impact of energy efficiency and other measures on CO_2 emissions. More specifically, it seeks to (1) increase the reliability of data for estimating emission reductions, (2) provide real-time data so the mid-course corrections can be made, (3) introduce consistency and transparency across project types and reporters, and (4) enhance the credibility of the projects with stakeholders.[29]

LEGISLATIVE PROPOSALS

In the 106th Congress, three types of legislation have been introduced that would support or otherwise affect the capacity for energy efficiency measures to curb global climate change. One category is direct appropriations for energy efficiency programs at DOE, EPA, and other agencies, which determine the range and magnitude of research, development, and implementation activities. The Clinton Administration's CCTI has sought major increases in spending for these energy efficiency programs as a strategy for curbing climate change. For FY2000, the appropriations for DOE energy efficiency programs (P.L. 106-113) supported some of the CCTI-requested increases. However, the appropriations for EPA energy efficiency programs (P.L. 106-74) did not fund any of the Administration's CCTI-requested increases.

A second category of legislation addresses the role of energy efficiency in curbing climate change by providing tax incentives for energy efficiency measures.[30] This category has included tax credits for homes, cars, and equipment that meet energy efficiency standards.

A third category of legislation focuses on policies to incorporate energy efficiency provisions in proposals to restructure the electric power industry. Since 1994, the emergency of state policies for electricity restructuring has dampened state and utility energy programs.[31] Some voice concern that a federal policy to restructure the electric industry could further cut back energy efficiency and other "clean" energy programs. However, some states' restructuring policies incorporate a public benefits fund (PBF) or other policy mechanisms to support energy efficiency. Recognizing this concern, some

---

[28] For more information, go to the website at [http://www.ipmvp.org/].

[29] Vine, Edward and Sathaye, Jayant. The Impact of Climate Change on the Conduct of Evaluation: The Establishment of New Evaluation Guidelines. In International Energy Program Evaluation Conference. *Evaluation in Transition: Working in a Competitive Industry Environment.* Ninth International Conference. Evanston, IL, August 1999. p. 435-446.

[30] The Clinton Administration's CCTI included tax incentives for energy efficiency.

[31] In particular, as the number of state restructuring policy proposals grew from 1995 through 1998, utility funding for demand-side management programs dropped sharply.

federal electricity restructuring bills have included provisions to support energy efficiency, including the creation of a PBF to support energy efficiency, a tax credit for combined heat and power (CHP) facilities, and other measures.

Table 1. Projected Energy Efficiency Contribution to Year 2010 Carbon Reduction Target (MMTC, million metric tons of carbon)

A. Include Carbon Sinks: 3% reduction from 1990 level					Change Relative to BAU		Change Relative to 1990		% of Kyoto Goal
		1990 Base	2010 BAU	2010 Policy	MM TC	%	MM TC	%	%
Five-Lab*	2010 BAU	1346	1730	1730	—	—	384	29%	—
	2010a	1346	1730	1604	126	7%	258	19%	30%
	2010b	1346	1730	1496	234	14%	150	11%	55%
	2010c	1346	1730	1336	394	23%	-10	-1%	93%
Kyoto (sinks)	—	1346	1730	1306	424	25%	-40	-3%	100%
EIA#	2010 BAU	1346	1791	—	—	—	445	33%	—
	2010	1346	1791	1466	325	18%	120	9%	67%
Kyoto (sinks)	—	1346	1791	1306	485	27%	-40	-3%	100%
B. No carbon sinks: 7% reduction from 1990 level									
Five-lab*	2010 BAU	1346	1730	1730	—	—	384	29%	—
	2010a	1346	1730	1604	126	7%	258	19%	26%
	2010b	1346	1730	1496	234	14%	150	11%	49%
	2010c	1346	1730	1336	394	23%	-10	-1%	82%
Kyoto (without sinks)	—	1346	1730	1252	478	28%	-94	-7%	100%
EIA	2010 BAU	1346	1791	—	—	—	445	33%	—
	2010	1346	1791	1466	325	18%	147	9%	56%
Kyoto (without sinks)	—	1346	1791	1252	539	30%	-94	-7%	100%

* *Five-lab Study* Scenarios (p. 1.14): (a) "energy efficiency," (b) high energy and low carbon, permit price = $25/ton carbon (HE/LC $25), (c) high energy and low carbon, permit price = $50/ton carbon (HE/LC $50).

Impacts of Kyoto Protocol Scenario 1990+9% (p. 146-150).

Source: DOE, *Scenarios of U.S. Carbon Reductions*, p. 1.12-1.14; DOE Energy Information Administration (EIA). *Impacts of the Kyoto Protocol on U.S. Energy Markets and Economic Activity*. October 1998. p. 120-122, 146-150.

Table 2. DOE Five-Lab Study: Potential Carbon Reductions form Energy Efficiency in 2010 (million metric tons of carbon, MMTC)

Sector	Scenario			
	1997 AEO BAU	Efficiency	HE/LC $25.00	HE/LC $50
Efficiency				
Buildings	—	25	42	59
Industry	—	28	44	62
Transportation	—	61	74	87
Total, Efficiency	—	**114**	**160**	**208**
Other	—	12	74	186
Grand Total	—	**126**	**234**	**394**
Net Emissions	1,730	1,604	1,496	1,336
Reduction below 1997 AEO business-as-usual (BAU)	0%	7%	14%	23%

Source: DOE, *Scenarios of U.S. Carbon Reductions*, Table 1.4, p. 1.12.

Chapter 3

GLOBAL CLIMATE CHANGE: A SURVEY OF SCIENTIFIC RESEARCH AND POLICY REPORTS

Wayne A. Morrissey

INTRODUCTION

It is difficult to ascribe a singular event that might have encouraged the U.S. Government to begin a major program to investigate global climate change; rather, it might be described as a long succession of events. The idea that carbon dioxide from industrial production could trap heat in Earth's atmosphere was proffered a century age, in 1898, by Swedish physicist Svante Arrhenius. Following the 1957-1958 International Geophysical Year,[1] scientists within U.S. federal agencies participated extensively in scientific workshops, international conferences, and international scientific research that explored the nature of Earth's climate system and the role of carbon dioxide (CO_2) and other greenhouse gases believed to modify it. In 1965 the President's Science Advisory Committee issued a report, *Restoring the Quality of Our Environment*, that identified climate change and CO_2 buildup as deserving expanded monitoring and study.

Notable early research programs included the Global Atmospheric Research Program (GARP) and the World Climate Research Program (WCRP). Around 1977, the prospect of global climate change had emerged from lecture halls and academic conferences and had begun to be presented in United Nations (U.N.) sponsored international fora attended by U.S. scientists. Not long after, international experts involved in research on potential global warming from greenhouse gas emissions would be asked by policymakers to contribute their scientific findings to an incipient international policy debate on the validity of concerns that global average temperatures might increase because of human activities.

---

[1] The International Geophysical Year's focus was organized to improve scientific knowledge about the Earth and its physical systems. Also, it celebrated the 100th anniversary of the first international Geophysical Survey undertaken to validate spatial measurements of the Earth.

The National Climate Program Act (NCPA) of 1978 (P.L. 95-367, 15 USC § 2901 et seq.) marked a major milestone in establishing a federal interest in global climate change policy. It signaled the beginning of a national policy, a diplomatic role in international actions on potential global climate change, and enhanced scientific research on these issues. It mandated coordination of domestic programs in climate research, applications, and services through an independent National Climate Program Office authorized under § 2908, of the Act. The NCPA also coordinated U.S. government participation in climate research conducted under international auspices.

The first major international studies on global climate change requested by world decision makers were performed by three United Nations organizations: the U.N. Environment Programme (UNEP), the World Meteorological Organization (WMO), and the International Council of Scientific Unions (ICSU). Their first study, which addressed potential global policy considerations, was released for public review and adopted by the U.N. Secretary General at the First World Climate Conference (FWCC) in Geneva in 1979. For about 10 years after FWCC, discussions continued the science of global climate change was further studied by WMO and UNEP, laying foundation for negotiations on a possible convention/treaty.

In November 1988, at the request of U.N. members involved in climate change research, the U.N. Secretary General, acting on recommendations of WMO and UNEP for an assessment of the state of knowledge about climate change, created the Intergovernmental Panel on Climate Change (IPCC). The IPCC's charge was to establish an orderly process to ensure that research and impact assessment studies proceeded concurrently and that adequate scientific research would precede legal or regulatory activities. To aid in its participation in the IPCC assessment process, and to coordinate U.S. research activities, the United States established the U.S. Global Change Research Program in 1989, pursuant to the Global Change Research Act (P.L. 100-606, 15 USC § 2921 et seq.). In 1990, the IPCC completed the first of a series of periodic international scientific and policy assessments of global climate change. These assessments aimed to "review scientific knowledge about natural consensus of scientific evidence on climate change and resulting impact on natural and human systems from which policy options can be developed."[2]

In 1990, after analysis of the first IPCC assessments, the U.N. General Assembly established an International Negotiating Committee (INC) to begin a process leading to a treaty on climate change. In 1992, the U.N. Framework Convention on Climate Change (FCCC) was adopted. The United States, along with 152 other nations, agreed to an ultimate objective of "stabilizing atmospheric greenhouse gas concentrations at a level that would prevent dangerous anthropogenic interference with the climate system." The U.N. FCCC established a non-binding goal and policy framework for the industrialized countries to pursue various voluntary measures to limit their emissions of greenhouse gases to 1990 levels by the year 2000. In October 1993, President Clinton outlined a voluntary Climate Change Action Plan (CCAP) intended to work toward stabilizing U.S. emissions of greenhouse gases.

---

[2] National Climate Program Office, January 15, 1988.

Prominent scientists were soon projecting that global greenhouse gas emissions would continue to rise long after the fundamental, voluntary commitments under the U.N. FCCC might be satisfied. Since the FCCC's "entry into force" in March 1994, debate has continued about the adequacy of scientific knowledge to be able to predict future climate change and how to deal with its possible impacts if it were occurring. Also, there has been debate about whether all U.N. FCCC parties – and not only industrialized countries – should participate in activities aimed at protecting Earth's climate.[3] Congress has held numerous hearings to inquire about the robustness of scientific findings relating to climate change and on other reports by social scientists projecting potential impacts for the U.S. economy, human health, and in other areas. Many studies have recommended possible policy responses, such as mitigation and adaptation strategies, to prevent or adapt to possible climate change. Members of Congress have also elevated concern about the issue internationally through direct communication with world leaders. Some have participated in international negotiations. Congress has passed resolutions and legislation that would help to coordinate scientific research on climate change, and has collected information from U.N. scientific bodies, and other international governmental and non-governmental organizations concerned with climate change.

While the U.N. FCCC focused on voluntary actions to be taken by the year 2000 for long-term control of the total concentration of atmospheric greenhouse gases, subsequent attention focused on regulating and reducing greenhouse gas emissions after the year 2000. As early as 1995, when U.N. FCCC parties were advised that is was unlikely that the voluntary goals of that treaty would be met, a Conference of Parties (COP) authorized under FCCC began to consider legally binding measures to reduce greenhouse gas emissions and proposed to craft a protocol or some other legal instrument that would be binding on the industrialized and developing countries.

The Berlin Mandate, which was adopted at the first meeting of COP (COP-1), in 1995, proposed dealing with future climate change by strengthening existing commitments under FCCC. However, it also continued to exempt developing countries, who are parties, from any new binding commitments related to controls on greenhouse gas emissions. Shortly thereafter, the IPCC released its second assessment on climate change (IPCC-2) in December 1995. Despite debate on its scientific findings, the United Nations endorsed IPCC-2 as the basis and scientific guidelines for negotiations on further action to limit possible human alteration of the climate system. One of the main findings of IPCC-2 – and one that has attracted considerable debate – was that "the balance of scientific evidence suggested a discernible human impact on the climate system." At the conclusion of COP-1, a two year "Analysis and Assessment" phase was begun to consider the possible elements of a regulatory instrument to limit greenhouse gas emissions.

Successive negotiations by the Conference of Parties helped to forge a December 1997 accord, the U.N. Kyoto Protocol on Climate Change, an international treaty which, if it enters into force, would implement the first legally binding reduction of greenhouse

---

[3] For a more in-depth discussion of this period of negotiations, see CRS Report 96-699 SPR, *Global Climate Change: Adequacy of Commitments under the U.N. Framework Convention and the Berlin Mandate.*

gas emissions with the aim of stabilizing (if not reduction) atmospheric concentrations of these pollutants at some point in the future. Different countries would be bound by different levels of responsibility and compliance under the Protocol, but combined efforts required of industrialized countries, *alone*, would be expected to reduce global emissions of greenhouse gases by some 5% from 1990 levels by the year 2012.[4]

On November 12, 1998, the United States became the 60th country to sign the Kyoto Protocol despite protest from some in Congress. It has not yet been submitted to the Senate for advice and consent to ratification. Members of Congress have since introduced legislation, and Congress has included provisions in FY1999 and FY2000 appropriations bills to prohibit activities that would implement the Kyoto Protocol without prior Senate advice and consent to ratification.[5]

SCIENTIFIC DEBATE ON GLOBAL
CLIMATE CHANGE AND U.S. POLICY

Many uncertainties continue to surround the theory of "global warming." At the very core of the scientific debate has been over the extent to which human activities influence climate change. Another uncertainty is whether the potential impacts of climate change might be harmful or beneficial for humans, managed agriculture, and natural ecosystems. Some question the validity and reliability of the scientific data to date which have underpinned negotiations toward possible international cooperation on regulation of greenhouse gases suspected to be causing globally averaged warming. Others are convinced that actions must be taken as soon as possible to reduce potential effects of human-added gases released into the atmosphere since the beginning of the industrial era (c.a. 1850).

A number of seminal scientific studies were performed by international scientific institutions prior to and since the U.S. began to formulate a national policy on global climate change. Subsequently, the state of the knowledge about climate change has evolved and so have future projections of the potential impacts of climate change as demonstrated by a variety of computer models of Earth's climate.

Atmospheric concentrations of carbon dioxide (CO_2), the major greenhouse gas, have increased by about one-third over the past 100 years or so; over this period global temperature has averaged an increase of an estimated 0.5°C (0.9°F). A majority of state-of-the-art computerized general circulation models (GCMs), which approximate the Earth's climate, have projected a globally averaged warming in a range of 3 to 8 degrees Fahrenheit over the next 100 years, it greenhouse gases were to continue to accumulate in the atmosphere at the current rate. Prominent climate scientists have concluded that such a warming could shift temperature zones, rainfall patterns, and agricultural belts and, under certain scenarios cause sea level to rise and inundate low-lying coastal areas.

---

[4] For more details, see CRS Report 98-2, *Global Climate Change Treaty: The Kyoto Protocol*. Also see, CRS briefing book on *Global Climate Change*, [http://www.congress.gov/brbk/html/ebgcctop.html]

[5] See CRS Report 98-664 STM, *Global Climate Change: Congressional Concern About "Back Door" Implementation of the 1997 U.N. Kyoto Protocol*.

Global warming, they believe, could have far-reaching effects – some positive, some negative, depending on regional impacts – on natural resources; ecosystems; food and fiber production; energy supply, use, and distribution; transportation; land use; water supply and control; and human health. However, other scientists who are skeptical of the "global warming" theory debate the credibility of available data, claim it is insufficient for public decision makers, and criticize the results of mathematics – and physics-based climate models which use these data."[6]

Most scientists are confident that the increase in atmospheric concentrations of carbon dioxide (CO_2) since the industrial revolution are primarily from human activities, and many of these scientists also conclude that this increase could be leading to higher global average temperatures. Other scientists, however, argue that scientific proof of the link between greenhouse gases and warming is inconclusive, or even contradictory, and that many uncertainties remain about the nature and future direction of Earth's climate. In any event, concern is growing that increased CO_2 from human activities, such as the burning of fossil fuels, industrial production, deforestation, and certain land-use practices, along with increasing concentrations of other trace gases, including chlorofluorocarbons, (CFCs), methane (CH_4), nitrous oxide (N_2O), hydro fluorocarbons (HFCSs), per fluorocarbons (PFCs) and sulfur hexafluoride (SF_6), may be changing the chemical composition and the physical dynamics of Earth's atmosphere, including how heat/energy is distributed among the land, ocean, atmosphere, and space.[7]

In its 1995 assessment of global climate change, the U.N. Intergovernmental Panel on Climate Change, (IPCC) reported a "discernible human impact on the climate system." Some critics argue, however, that the signal has not emerged clearly from the background noise of natural climate variability that has transpired over long time periods. Lead authors of the IPCC scientific working group countered that uncertainties were adequately addressed in Chapter 8 ("Detection of Climate Changes and Attribution of Causes") in the Science Working Group Report, and in the contents of other IPCC working group reports.

Skeptics have testified in congressional hearing that, upon reviewing the latest scientific literature about global climate change, they believe that 1) carbon dioxide may be 15% less potent in its ability as a greenhouse gas to warm the climate than previously thought, 2) methane has stabilized in the atmosphere and may be on the decline, calling into question the magnitude of its radiative contribution to climate change, and 3) global average temperatures have not risen at the rate or magnitude projected by GCMs. All of which, they claim, call into question the reliability of the computer climate models used to make projections of future warming and those that served as the basis for Kyoto Protocol negotiations.[8]

---

[6] For a more in-depth discussion, see "Policy Forum: Uncertainties in Projections of Human-Caused Climate Warming," by J.D. Mahlman. *Science*, vol. 278, Nov. 21, 1997; 1416-1417.

[7] See CRS Issue Brief IB89005, *Global Climate Change*.

[8] Testimony of Patrick Michaels, Professor of Environmental Sciences, University of Virginia and Senior Fellow in Environmental Studies at the Cato Institute to House Commerce Committee, Subcommittee on Small Business, July 27, 1998.

Skeptics have also challenged some scientists' interpretations that recent episodic weather events, which seem more extreme in nature, are indicative of long-term climate change. The Clinton Administration has been criticized for suggesting the possibility that recent floods, forest fires, rather severe weather and certain other weather anomalies could relate to a warming of the climate.

Natural variability of climate (or climate fluctuation) is large enough that statistically even the record-setting warmth and severe weather events of the 1980s and 1990s cannot be attributed entirely to human activities. In some cases, connections between inter annual and inter decadal climate fluctuations such as *El Nino* and *La Nina* and seasonal patterns of severe weather events are just beginning to be recognized, largely because of an improved ability to observe the nature, frequency, and severity of atmospheric and oceanic phenomena. This notwithstanding, singular extreme weather events have focused public attention on possible outcomes of potential long-term climate change and the need for a better understanding of regional climates. Consequently, it appears that many scientific questions about the nature of climate change and its relationship to regional weather patterns remain to be answered.

National Oceanic and Atmospheric Administration's (NOAA) researchers have reported that the 12 warmest years, in terms of global averages, since historical records have been kept (since about 1780) occurred during the past two decades, with 1990, 1998, and 1999 among the warmest. At least some of this warming, they concluded, is human-induced, because of the rate of change observed since post-industrial times. On the other hand, satellite instruments – which measure radiative properties of certain gases from which average temperatures of the atmosphere in a deep column above the surface may be deduced – have not demonstrated any significant temperature rise at upper levels of the atmosphere over the past 20 years. A recent report sheds some light on this debate. (See National Research Council, below.) Arguments such as this have cast some doubt as to whether historic temperature data – among other indicators of climate change, such as ice core sampling and tree ring data, for example – are a reliable way to estimate future atmospheric temperature. Scientists have also debated whether recent emergence of tropical diseases in the mid-latitudes and apparent biological changes manifested in certain species of flora and fauna are signaling that Earth's climate is warming on average. In efforts to address many of these unresolved issues, a third IPCC assessment of global climatic change is expected late in the year 2000, and will likely influence future negotiations on global climate change.

MAJOR REPORTS BY INTERNATIONAL SCIENTIFIC INSTITUTIONS

The following selection of reports identifies major studies from an international perspective that have contributed to U.S. policy debates on global climate change. They are generally listed in chronological order and are attributed to the scientific institution responsible for the work. These include some of the earliest studies on climate change as it became a science policy issue for the United States and in international affairs. These represent a progression of views during the ten-year period between the first World

Climate Conference in 1979 and the eventual establishment of the U.S. Global Change Research Program in 1989, which defined a U.S. scientific research framework for global climate change. Those listed are selected based upon endorsement, involvement, or membership of U.S. government scientists in the WMO, UNEP, and ICSU. For additional studies by individuals, public interest groups, and others, see the CRS global climate change electronic briefing book web site. [http://www.congress.gov/brbk/html/ebgcc.html].

U.N. World Meteorological Organization (WMO)

Three important reports were issued as a result of activities stemming from the 1979 World Meteorological Organization (WMO) First World Climate Conference (FWCC), co-sponsored by UNEP and ICSU, which resulted in the creation of a U.N. World Climate Research Program. These organizations began to consider what might be possible issues of concern for world decision makers, and reviewed administrative responsibilities of the WMO as far as its role in international research on climate change. Many other reports by WMO have followed, focusing on more specific aspects of global climate change research and its potential role in informing public decision makers. In addition, other important reports by WMO's plenary body and WCRP have been issued. The three reports are:

- *Proceeding of the World Conference: A Conference of Experts on Climate and Mankind:* Geneva, Switzerland, February 12-23, 1979. Secretariat of WMO, Geneva: 1979. W MO-No.537.
- *Report of the International Conference on the Assessment of the Role of Carbon Dioxide and of Other Greenhouse Gases in Climate Variations and Associated Impacts:* Villach, Austria, October 9-15, 1985. World Climate Research Program, WMO/UNEP/ICSU, Geneva: 1986. WMO-No. 661.
- *World Climate Program Impact Studies: Developing Policies for Responding to Climatic Change; a Summary of the Recommendations of the Workshops Held in Villach* (28 September-2 October 1987) *and Bellagio* (9-11 November 1987) Beijer Institute, Stockholm: April 1988.

Two other WMO publications are published on its Internet website [http://www.unep.org/ipcc/qa/cover.html]. These discuss contemporary aspects of the scientific debate on global climate change, and include:
"Common Questions About Climate Change". This report is undated. It poses 10 of the most commonly asked questions about climate change, including whether the Earth has warmed, which human activities might be contributing to climate change, what further climate changes are expected to occur, and what effects these changes may have on humans and the environment, and suggests possible answers for those questions.
"Scientific Assessment of Stratospheric Ozone Depletion". WMO Global Ozone Research and Monitoring Project-Report Series. *An Assessment of our Understanding of*

the Processes Controlling its Present Distribution and Change. This report is prepared every 3 years in cooperation with the UNEP, the U.S. National Aeronautics and Space Administration, Federal Aviation Administration (FAA), National Oceanic and Atmospheric Administration (NOAA), the Commission of the European Communities and, beginning in 1989, the intergovernmental Alternative Fluorocarbon Environmental Acceptability Study (AFEAS) team. This report series offers insight into the potential role of some chlorofluorocarbons (CFCs) in stratospheric ozone depletion and global warming, as well as the potential climate effects of their approved replacements. This report acknowledged that, although regulated under the 1987 Montreal Protocol on Substances that Deplete the Ozone Layer for their ozone depleting effects, some these compounds have, in addition, a Total Environmental Warming Potential (TEWP) while others may produce secondary effects on the climate by cooling the stratosphere.

International Council of Scientific Unions (ICSU)

International Geosphere-Biosphere Program (IGBP): A Study of Global Change. Global change report series, report no. 4: A Plan for Action. International Council of Scientific Unions (ICSU), Special Committee for the IGBP. Stockholm, Sweden: August 1988. This report, prepared for discussion at the First Meeting of the Scientific Advisory Council for the IGBP, Stockholm, Sweden, October 24-28, 1988, laid out a framework for "(1) Documenting and predicting global change; (2) Observing and improving our understanding of dominant forcing functions; (3) Improving our understanding of interactive phenomena in the total Earth system; (4) Assessing the effects of global change that will cause large-scale important modifications in the availability of renewable and non-renewable resources." Working groups also evaluated current and projected research capacity in four areas: (1) global geosphere-biosphere modeling; (2) data and information systems; (3) techniques for extracting environmental data of the past; and (4) geosphere-biosphere observatories.

The U.N. Commission on Environment and Development (UNCED)

Our Common Future. U.N. Commission on Environment and Development (UNCED). Geneva, Switzerland: 1987. The General Assembly of the United Nations called upon UNCED to propose long-term environmental strategies for achieving sustainable development by the year 2000 and beyond in ways that would promote greater co-operation among developing countries and between countries at different stages of economic and social development. As part of this report, the Commission stated that "The burning of fossil fuels puts into the atmosphere carbon dioxide, which is causing gradual global warming. This greenhouse effect may be early next century have increased average global temperatures enough to shift agricultural production areas, raise sea levels to flood coastal cities, and disrupt national economies."

The U.N. Intergovernmental Panel on Climate Change (IPCC)

Established by the U.N. in November 1988, and made up of prominent scientists from WMO and UNEP governing bodies and representatives of national climate change research programs, the IPCC was charged with performing the first internationally sponsored assessment of global climate change. Its members were requested by world governments to consider issues associated with the science, impacts, and possible response strategies to prepare for the possible onset of a greenhouse warming.

IPCC First Assessment Report: Overview. World Meteorological Organization and United Nations Environment Programme, Geneva: August 31, 1990. This report focused on the science of, impacts of, and responses to potential global climate change, and contained a policymakers' summary of the IPCC Special Committee on the Participation of Developing Countries. The overview brought together material from the four IPCC policy-makers summaries. It presented conclusions, proposed lines of possible actions (including suggestions of factors which might form the basis for negotiations) and outlined further work required for a more complete understanding of the problem of climate change resulting from human activities. This document summarized the three full reports of the working groups on Science (WG-1), Impacts (WG-2), and Responses (WG-3). Also, the Overview Report was an opportunity for additional technical assessment by experts of those governments that could not participate in the three Working Groups of IPCC, but was not one which reflected individual government's positions.

IPCC Supplement: Radiative Forcing of Greenhouse Gases. IPCC, Geneva: February 1992. This report presented an overview and findings of a meeting held in Guangzhou, China, to assess latest scientific data on global climate change, including new estimates of the indirect global warming potential of some greenhouse gases, and to update or confirm findings originally put forth in the three IPCC Working Group Reports completed in the summer of 1990.

IPCC Second Assessment Synthesis of Scientific-Technical Information Relevant to Interpreting Article 2 of the U.N. Framework Convention on Climate Change and Summaries for Policy-makers of Working Groups I, II, and III Reports of the IPCC. IPCC Secretariat: Geneva: December 1995. this report presents a review of the role of developing countries in greenhouse gas emissions reduction and reporting requirements, and a summary of major findings of the three working group reports: 1) science, 2) impacts, adaptation, and mitigation, and 3) economic and social dimensions of climate change. The chairman of the science working group released a controversial claim in a policy-makers summary stating that "The balance of scientific evidence suggests a discernible human influence on climate."

IPCC Technical Papers (1-4). IPCC. Geneva: 1997. Four technical papers were produced by IPCC working groups at the request of the Ad Hoc Group on the Berlin Mandate. Papers delved more in depth into a number of technical issues surrounding potential climate change agreements, including elements of a possible international regulatory framework (i.e., in support of a possible future protocol), and issues not resolved by the IPCC's 1995 assessment. The 4 papers include:

- *Technologies, Policies and Measures for Mitigating Climate Change*, November 1996;
- *An Introduction to Simple Climate Models Used in the IPCC Second Assessment Report*, February 1997;
- *Stabilization of Atmospheric Greenhouse Gases: Physical, Biological, and Socioeconomic Implications*, February 1997; and
- *Implications of Proposed CO$_2$ Emissions Limitations*, October 1997.

IPCC Special Report, The Regional Impacts of Climate Change: An Assessment of Vulnerability. IPCC, Geneva: November 1997. To expand upon the work of the IPCC's Second Working Group, the report consisted of vulnerability assessments for 10 regions that comprise the Earth's entire land surface and adjoining coastal seas. It also included annexes that provide information about climate observations, climate projections, vegetation distribution projections and socioeconomic trends.

MAJOR REPORTS BY U.S. SCIENCE INSTITUTIONS AND FEDERAL AGENCIES

Official U.S. scientific organizations such as the National Research Council and federal government agencies have been involved in scientific research on global climate change for many years now. Some of the latter have also been mandated by Congress under U.S. law to produce reports that present a snapshot of U.S. policy on climate change. The U.S. Global Change Research Program (USGCRP) began in 1989. Today, representatives from nine federal agencies contribute focused scientific research to USGCRP and, along with a few additional agencies, contribute, indirectly to global climate change research efforts through various intramural scientific research programs. The following science institutions and agencies are listed in chronological order of their involvement with the U.S. global climate change research and policy debate.

U.S. National Research Council (NRC)

The National Research Council (NRC) of the National Academy of Sciences, a federally chartered institution of scientists, has played an important role in organizing U.S. efforts in international studies on the state of the knowledge about global climate change. The NRC Global Change Committee, for example, serves as the U.S. representative for the International Geosphere-Biosphere Program – a global change study – sponsored by ICSU. NRC has also advised federal agencies on the broad range of multidisciplinary earth systems studies being undertaken to garner knowledge about global climate change, and has served as a major forum for communications between scientists and policy-makers on these issues.

Early NRC reports (pre-1983) focused on scientific research related to the issue, and summarized finds on the state-of-knowledge about global climate change and the role of CO_2. These were of a technical nature and reviewed scientific findings, not policy questions. The significance of these reports was that they represented the opinion of an internationally respected scientific institution, and reported the consensus of many of the world's leading scientific experts on climate change that carbon dioxide in the atmosphere had the potential to pose a significant threat to the environment by causing a warming of the Earth's average global temperature. A few examples of these are:

- *Studies in Geophysics: Energy and Climate.* National Academy Press, Washington, DC: 1977. Geophysics Study Committee, Geophysics Research Board.
- *Carbon Dioxide and Climate: a Scientific Assessment.* National Academy Press, Washington, DC: 1979. Report of an Ad Hoc Study Group on Carbon Dioxide and Climate, Woods Hole, MA, July 23-27, 1979.
- *Carbon Dioxide and Climate*: A Second Assessment: *Report of the CO$_2$ Climate Review Panel.* Climate Research Committee, Board on Atmospheric Sciences and the Carbon Dioxide Assessment Committee on the Climate Change Board.

Subsequent NRC reports began to include scientific findings of interest to public policymakers, and also addressed science policy concerns such as prioritizing a research framework, coordinating federal agency activities, allocating research funding for global change research programs, and reviewing strengths and shortcomings in the federal science infrastructure that operates such programs.

Changing Climate: Report of the Carbon Dioxide Assessment Committee: National Academy Press, Washington, DC: 1983. The Commission on Physical Sciences, Mathematics and Resources, Board on Atmospheric Sciences and Climate prepared this report in response to the Energy Security Act of 1980 (P.L. 96-294, 41 USC § 8911) to assess the potential impacts of the buildup of CO_2 in the atmosphere from the full-scale production of synthetic fuels. Authors of the report state, "Our stance is conservative: we believe there is reason for caution, not panic. Since understanding and proof of what is happening to climate as a result of practices that load the atmosphere with CO_2 may come too late to allow for corrective action, we may not be able to wait to make certain there is a best course." This is, perhaps, one of the first "policy" documents on the issue of global climate change contributed to by scientists, and one that was required by Congress, under P.L. 96-294, Title VII, subtitle B-Carbon Dioxide Study.

Policy Implications of Greenhouse Warming: Mitigation, Adaptation, Science, and Synthesis Documents. National Academy Press, Washington, DC: 1991. The Synthesis Panel. The House Committee on Appropriations called for and NAS study on the potential societal impacts of global climate change. (H.Rept. 100-701: 26). This was funded by an EPA grant approved under the HUD-Independent Agencies Appropriations Act of 1989 (P.L. 100-404, 42 USC § 13381). The Committee on Science Engineering and Public Policy of National Academy of Sciences undertook this study, popularly known as the "COSEPUP study." Three reports were prepared and released in the following order: *Adaptation* (August 1991); *Mitigation* (June 1991): and *Science* (September 1991). However, the *Synthesis Panel Report*, which summarized the findings of all three COSEPUP reports, was published first in April 1991, and proposed least cost strategies for reducing U.S. greenhouse gas emissions 10%-40% of 1990 levels by the year 2000. The panel concluded that some greenhouse gas emission reductions could be realized at a net savings if appropriate policies were implemented.

A Decade of International Climate Research: The First Ten Years of the World Climate Research Program. National Academy Press, Washington, DC: 1992. Climate Research Committee, Board on Atmospheric Science and Climate. One component of the WMO, the World Climate Research Program (WCRP), is reviewed by NAS at the request of the U.S. National Climate Program Office on behalf of federal agencies which support climate change research, observation systems, and services. The report findings state, "The WCRP, a framework for cooperation that has been active for a decade, has made measurable progress in leading all nations to a better understanding of climate." The Committee assessed the principle achievements and shortcomings of WCRP, and included conclusions about and recommendations for future direction.

Overview: Global Environmental Change: Research Pathways for the Next Decade. National Academy Press, Washington, DC: 1998. Committee on Global Change Research, Board on Sustainable Development, Policy Division. Participants noted that deliberations over the Kyoto Protocol set "environmental goals, which would affect the science priorities as well as economic paths in the coming century, and that scientists needed to create an intellectual framework to hone the questions that need immediate attention, to separate the vital from the interesting, an to preserve basic research for discovery of the unexpected." The Committee provided guidance on such a framework and clarified pathways for planning future U.S. research on global climate change. The report summarized background, findings, and recommendations of the Committee and reviewed research over the past decade, especially that of the U.S. Global Change Research Program (USGCRP). This initial charge to assess performance of the USGCRP would serve as a basis for a future report that would: 1) articulate the central scientific issues posed by global environmental change; 2) state the key scientific questions which must be addressed by USGCRP; and 3) identify the scientific programs, observational efforts, modeling strategies, and synthesis activities needed to attack these scientific questions. The Committee called for a revitalization of USGCRP and stressed the importance of U.S. leadership in supporting global change research.

Decade-to-Century-Scale Climate Variability and Change: A Science Strategy. National Academy Press, Washington, DC: 1998. Commission on Geosciences, Environment, and Resources. Board on Atmospheric Sciences and Climate and Climate Change Committees. "Dec-Cen" Panel on Climate Variability on Decade-to Century Time Scales. Panel members reported, "In 1990, the Intergovernmental Panel on Climate Change (IPCC) released its monumental first scientific assessment on climate change...One significant gap involved our meager understanding and documentation of natural variability in Earth's climate system which provides a context for evaluating the significance of human-induced changes." This report formulated a research strategy and presented those scientific issues and infrastructure considerations required to most effectively advance understanding of climate variability and change on decade-to-century time scales. It also emphasized steps necessary to more confidently predict future climate conditions and detect climate change as part of a holistic research perspective, which the panel believed is required to address this issue.

The Atmospheric Sciences: Entering the 21st Century. National Academy Press, Washington, DC: 1998. Commission on Geosciences, Environment, and Resources. Board on Atmospheric Sciences and Climate. This report set forth recommendations intended to strengthen atmospheric sciences and provide climate prediction services intended to benefit for the nation. Board members concluded, "It [the report] is thus intended for those who share the responsibility for maintaining the pace of improvement in the atmospheric sciences, including leaders and policy makers in the public sector, such as legislators and executives of the relevant federal agencies; decision makers in the private sector of the atmospheric sciences; executives of other economic endeavors whose activities are dependent on atmospheric information, and of course university departments that include atmospheric science."

Adequacy of Climate Observing Systems. National Academy Press, Washington, DC: 1999. Commission on Geosciences, Environment, and Resources. Board on Atmospheric Sciences and Climate. This report discussed how instrumentation, observing practices, processing algorithms, and data archive methods used by scientists may profoundly affect the understanding of climate change. The Board assess whether scientists are making the measurements, collecting the data, and making it available in a way that would enable contemporary and future scientists to effectively increase understanding of natural and human-induced climate change. The report concluded that this was not the case, and illuminated the importance of multi-decadal climate monitoring and recommended strategies to achieve those goals.

Reconciling Observations of Global Temperature Change. National Academy Press, Washington, DC: January 2000. National Research Council, Board on Atmospheric Sciences and Climate Panel. This report discusses whether the observed surface warming of the Earth (over the past 20 years) is real or a product of unreliable and inconsistent data. Also, the report attempts to resolve disparities between temperature trends measured at the surface and upper air temperature trends from satellite data which skeptics have claimed may invalidate the results of general circulation models (GCMs) that have forecasted future climate change. Critics of GCMs point out that results of model runs demonstrate a homogeneous warming throughout all the layers of the Earth's atmosphere. Panel scientists believe that there may be a systematic disconnect between the upper and near surface atmosphere and have cited physical processes, which may have an unique impact on the upper atmosphere that are not currently accounted for in GCMs. The Panel reported that because of scientific uncertainties the difference in temperature trends cannot be explained. This, they concluded, was because of the paucity of surface and radiosonde data for some geographic locations, and the lack of consistent, long-term monitoring of the upper atmosphere.

U.S. Department of Energy (DOE)

Carbon Dioxide Research: State-of-the-Art Report Series. Office of Energy Research and Office of Basic Energy Sciences, Carbon Dioxide Research Division, Washington, DC. DOE reported that the enormity and diversity of the problem of coordination of multi-disciplinary research on carbon dioxide made it difficult to: define the problem; develop strategies for solving the problem; and establish communication and cooperation among the researchers working on different facts of the problem. DOE also recognized that the compilation, integration, interpretation; and dissemination of information were especially important. To improve communication between scientists and public decision makers, DOE prepared four State of the Art reports:

- Atmospheric Carbon Dioxide and the Global Carbon Cycle, February 1986;
- Direct Effects of Increasing Carbon Dioxide on Vegetation, March 1986;
- Detecting the Climatic Effects of Increasing Carbon Dioxide, February 1986;
- Projecting the Climatic Effects of Increasing Carbon Dioxide, April 1986.

Two additional reports, completed earlier, were later added to the series:

- Characterization of Information Requirements for Studies of CO_2 Effects:
- Water Resources, Agriculture, Fisheries, Forests, and Human Health, October 1985; and Glaciers, Ice Sheets, and Sea Level: Effects of a CO_2-Induced Climatic Change, 1984.

A Compendium of Options for Government Policy to Encourage Private Sector Responses to Potential Climate Change. U.S. Dept. of Energy, Washington, DC, October 1989. This is a compendium of generic policy instruments and specific policy options available to the U.S. government if it chooses to require significant private sector efforts to prevent, mitigate, or adapt to climate change. Authors of the report pointed out, "The selection of any particular package...is a largely political choice of preferred means to achieve the overall policy goal."

Interim Report of the National Energy Strategy: A Compilation of Public Comments. U.S. Dept. of Energy, Washington, DC: April 1990. The executive summary of the document stated that "Consistent with the President's directive to build national consensus, we have begun the task of developing a National Energy Strategy by opening a dialogue with the American people. We (DOE) have held fifteen public hearings in many areas of the country, several co-chaired by Cabinet Secretaries from other Federal agencies. More than 375 witnesses representing 43 States have contributed to several thousand pages of testimony. Further, our efforts to seek input from State and local governments, consumer organizations, business, industry, and recognized representatives of diverse points of view have resulted in more than 1,000 written submissions. The purpose of the Interim document is to convey the results of this public dialogue [on a National energy strategy]...The comments received are organized on the basis of presented public concerns, publicly identified goals, publicly identified obstacles to achieving those goals, and publicly suggested options for action to remove or overcome the obstacles."

Scenarios of U.S. Carbon Reductions: Potential Impacts of Energy Technologies by 2010 and Beyond, "The Five-Lab Study". U.S. Dept. of Energy. Washington, DC, September 1997. The report analyzed some options for using cost-effective, high efficiency energy technologies and other low carbon technologies to curb greenhouse gas emissions. It also estimated the potential cost per ton for carbon reduction required to stabilize U.S. emissions at 1990 levels by 2010, through implementation of such technologies. The 5-lab study also concluded that all emission-reduction scenarios that were modeled could be achieved at low or no net costs.

Impacts of the Kyoto Protocol on U.S. Energy Markets and Economic Activity. U.S. Dept. of Energy, Energy Information Administration (EIA), Washington, DC, October 1998. The analysis in this report was undertaken at the request of the U.S. House of Representatives Committee on Science on March 3, 1998,[9] to analyze potential economic impacts of the Kyoto Protocol, by focusing on different scenarios for U.S. energy use and prices and the economy in the years 2008 to 2012. This report was prepared as a critique of the DOE "5-Lab Study," described above, and basically disagreed with the former's economic conclusions. Authors noted, "the report was prepared with sensitivities evaluating key uncertainties [such as]: U.S. economic growth, the cost and performance of energy-using technologies, and the possible construction of new nuclear power plants."[10]

Emissions of Greenhouse Gases in the United States. DOE, Energy Information Administration, Office of Integrated Analysis and Forecasting. DOE/EIA-0573(98), Washington, DC: October 1999. Annual report series begun in 1996, reflecting 1995 U.S. emissions. These documents report U.S. aggregate greenhouse gas emissions based upon reporting consequent to § 1605 of the Energy Policy Act of 1992 (P.L. 102-486, 42 USC § 13385). They also project future emissions based upon projected energy demand. Latest estimates of emissions for carbon dioxide, methane, nitrous oxide, and other greenhouse gases are included.

National Aeronautics and Space Administration (NASA)

Earth Systems Science: A Program for Global Change. National Aeronautics and Space Administration, Earth System Sciences Committee, NASA Advisory Council, Washington, DC: January 1988. This extensive study proposed near-term (1987-1995) and long-term (1995 and beyond) recommendations for: 1) sustained, long-term measurements of global variables; 2) a fundamental description of the Earth and its history; 3) research foci and process studies; 4) development of Earth system models; 5)

---

[9] See Appendix D of the cited report: "Letters from the House Committee on Science."
[10] The legislation establishing EIA in 1977 vested the organization with an element of statutory independence. EIA does not take positions on policy questions, and does no purport to represent the official position of the Department of Energy or the Administration.

an automated information system/clearinghouse for Earth system science; 6) coordination of federal agencies activities; and 7) enhanced international cooperation.

National Oceanic and Atmospheric Administration (NOAA)

Reports to the Nation on Our Changing Planet: Our Changing Climate. Dept. of Commerce, National Oceanic and Atmospheric Administration, Boulder, CO: Fall 1997. This report, prepared under a grant to University Corporation for Atmospheric Research, succinctly reviewed the state of the knowledge of climate change as of its publication and explored the role or natural and possibly human-induced changes. Authors stated that, "We have entered an era when actions by humanity may have as much influence on Earth's climate as the natural processes that have driven climate change in the past. Our future climate will be partly of our own making."

National Science Foundation (NSF)

Global Change. (NSF) *Mosaic* 19, ¾, Fall/Winter 1998: entire issue. This special double issue outlined a potential framework for international climate change research and introduced the major players in science policy in the global change community. It discussed a potential role for the U.S. in the International Geosphere Biosphere Program (IGBP). The editor of this issue states, "The planners of the worldwide effort to untangle the processes that lead to global change have seized the moment by producing an outline: now its up to the scientific communities of many nations and many disciplines to fill in the blanks."

White House Office of Science and Technology (OSTP)

Our Changing Planet: The U.S. Global Change Research Program. U.S. Office of Science and Technology Policy, Committee on Environment and Natural Resources, Washington, DC: 1989. Originally issued as *A U.S. Strategy for Global Change Research*, a report by the Committee on Earth Sciences, to accompany the President's FY 1990 Budget, with expectations that this document would be released annually thereafter (pursuant to P.L. 101-606, 15 USC § 2921 et seq.). This report was followed by a formal FY1990 "research plan" which looked forward 10 years. Only one more formal research plan was released for FY1991. Five-year assessments were called for thereafter. The FY1990 report stated that, "The purpose of this document is to provide an initial research strategy to guide planning and conduct of the U.S. Global Climate Change Research Program (USGCRP)."

These comprehensive research plans presented details of the USGCRP, evaluated how well the current activities addressed the key scientific questions and program goals, identified the gaps in knowledge, prioritized among research needs, and defined individual federal agency roles. They were developed in close collaboration with other national and international planning groups and activities, including the National Academy of Sciences and the International Geosphere-Biosphere Program, and took into account programs outlined in the 5-year plan of the National Climate Program. After FY1992, the research plan was integrated into the *Our Changing Planet (OCP)*" budget document, which has been published annually through FY2001. For FY2000, the OCP contained a section entitled "Perspectives for the USGCRP for the decade ahead, preparing the agenda for the 21st century."

U.S. Environmental Protection Agency (EPA)

The Potential Effects of Climate Change on the United States. U.S. Environmental Protection Agency, Office of Policy, Planning, and Evaluation, Washington, DC: December 1989. The Continuing Resolution Authority for FY1987 (P.L. 99-591, 15 USC § 2901, note) mandated that EPA conduct a study on the greenhouse effect and to prepare two reports which focus on the health and environmental effects of climate change. The first report focused on the potential health and environmental effects of climate change including, but not limited to, the potential impacts on agriculture, forests, wetlands, human health, rivers, lakes, estuaries as well as societal impacts, and it was structured to address regional impacts of climate change on the Southeast, the Great Plains, California, and the Great Lakes.

Policy Options for Stabilizing Global Climate Change. U.S. Environmental Protection Agency, Office of Policy, Planning, and Evaluation, Washington, DC: December 1990. Second of two reports on the greenhouse effect mandated by Congress in P.L. 99-591. The report presented, "A comprehensive and global approach," covering all sectors and all greenhouse gases, in the analysis of policy options, from energy efficiency to new methods of rice cultivation, and presented possible future scenarios of greenhouse gas emissions up to the year 2100, with different levels of policy response, and other independent factors, such as domestic economic performance.

U.S. Efforts to Address Global Climate Change: Report to Congress and Appendices. U.S. Environmental Protection Agency, Office of Policy, Planning and Evaluation and U.S. Department of State, Washington, DC: February 1991. Mandated by Congress in the Global Climate Protection Act of the *Foreign Relations Authorization Act, Fiscal Years 1988 and 1989* (P.L. 100-204, § 1103, Title XI; Global Climate Protection – Global Climate Protection Act of 1987). Section 1103 expressed certain congressional findings regarding global climate protection, including the following: (1) there is evidence that manmade pollution may be producing a long-term and substantial increase in the average temperature on the surface of the Earth, a phenomenon known as the "greenhouse" effect; and (2) vigorous research is required in order to prevent such pollution from altering the global climate, and affecting agriculture and habitability over large portions of the Earth's surface within the next century. Also, in Title XI of this Ace, the President, through EPA, was directed to develop and propose a coordinated national policy on global climate change; and to direct the Secretary of State (hereafter Secretary) to coordinate such U.S. policy in the international arena. The Secretary and the

Administrator of the EPA (hereafter Administrator) were directed within 24 months after enactment of this Act to jointly report to the appropriate congressional committees an analysis, description, and strategy of the United States with respect to the greenhouse effect and its potential role in global climate change. Congress had also urged the Secretary of State to promote an International Year of Global Climate Protection (IYGCP), and encouraged the President to accord the problem of climate protection a high priority on the agenda of U.S.-Soviet relations. The President did non endorse U.S. involvement in an IYGCP. The resulting report identified U.S. efforts to address potential climate change, and recommended that U.S. policy should seek to: (1) increase worldwide understanding of the greenhouse effect and its consequences: (2) foster cooperation among nations to coordinate research efforts with respect to such effect; and (3) identify technologies and activities that limit mankind's adverse effect on the global climate.

Inventory of U.S. Greenhouse Gas Emissions and Sinks: 1990-1996. U.S. Environmental Protection Agency, Office of Policy, Planning, and Evaluation, Washington, DC: March 1998. EPA236-R-98-006. This annually published report summarizes the latest information of U.S. greenhouse gas emissions trends from 1990, for emissions sources related to energy consumption, land-use change and forestry (CO_2), hydro fluorocarbons, (HFCs), per fluorocarbons (PFCs), sulfur hexafluoride (SF_6) and selected methane (CH_4) sources.

U.S. Department of State (DOS)

Climate Action Report: 1997 Submission of the United States of America Under the U.N. Framework Convention on Climate Change. U.S. Dept. of State, Washington, DC: 1996. DOS Pub. 10496. The report stated, "This document has been developed using the methodologies and format agreed to at the First Conference of Parties to the FCCC, and modified by the second meeting of the Conference of Parties and by sessions of the Convention's Subsidiary Body on Scientific and Technological Advice and the Subsidiary Body on Implementation." Also, in the report the United States stated that is assumes that this communication, like those of other countries – and like the proceeding U.S. communication (December 1992) – would be subject to a thorough review and discussed in the evaluation process for the Parties of Convention. The authors noted that, "Even though the measures listed in this report are not expected to reduce U.S. emissions below 1990 levels by the year 2000, the United States believes that many of the climate change actions included in this report upon being implemented have been successful at reducing emissions, send valuable signals to the private sector, and may be appropriate models for other countries."

U.S. Executive Office of the President (EOP)

America's Climate Change Strategy: An Action Agenda. By President George Bush, Washington, DC: February 1991. This report highlighted comprehensive actions to be taken to mitigate or adapt to potential climate change, and featured possible actions that "make sense," as well as reviewed actions already being undertaken by the Bush Administration to address global climate change, such as energy efficiency improvements, reforestation pursuant to the "America the Beautiful Program," under the 1990 Farm Bill (P.L. 101-624, Title XXIII, 16 USC § 2101), and through proposed reductions of CFCs, under Title VI of the Clean Air Act Amendments of 1990 (P.L. 101-549, 42 USC § 7671a et seq.).

Climate Change Action Plan. By President William J. Clinton and Vice President Albert T. Gore, Jr., Washington, DC: October 1993. On October 19, 1993, President Clinton released his *Administration's Climate Change Action Plan* (CCAP), which featured domestic measures that might be taken to attain the goal of greenhouse gas emissions stabilization as outlined under the terms of the U.N. Framework Convention on Climate Change, which were comparable to the President's own emissions goals. The CCAP has relied on a comprehensive suite of voluntary actions by industry, utilities, and other large-scale energy users. It also promoted energy-efficiency upgrades through devising new building codes in residential and commercial sectors, as well as other energy efficiency improvement in generic energy-generating capacity and energy consumption. Large-scale tree planting and forest reserves were also encouraged to enhance sinks for atmospheric carbon dioxide and conserve energy. Other provisions of the plan called for increased utilization of hydroelectric power sources, including upgrading existing facilities; encouraged use of public transportation; regulated methane release in land fills and capture of waste methane to be utilized as a fuel source. In addition the president called for controls on nitrous oxide, and on hydro chlorofluorocarbon (HCFC) byproducts believed to be contributing to global warming.

The Kyoto Protocol and the President's Policies to Address Climate Change: Administration Economic Analysis. White House Council of Economic Advisors (CEA) and others. Washington, DC, July 1998. The purpose of the report was to examine costs and benefits of taking action to mitigate the threat of global warming, and in particular, the costs of complying with the emissions reduction target for the United States set forth in the Kyoto Protocol on Climate Change, negotiated in December 1997. The report concluded that, "With the flexibility mechanisms included in the treaty, and by pursuing strong domestic policies, the United States can reach its Kyoto target at relatively modest cost. Moreover, the benefits of mitigating climate change are likely to be substantial."

U.S. Department of Justice (DOJ)

A Comprehensive Approach to Addressing Potential Climate Change. U.S. Dept. of Justice, Environment and Natural Resources Division, Task Force on the Comprehensive Approach to Climate Change, Washington, DC: February 1991. The work of this group was the model for President's Bush's *Action Agenda*, described above. The Task Force was created as a federal interagency effort with representatives from the President's Council of Economic Advisors, Council on Science and Technology, Council on Environmental Quality, White House Office of Policy Development and Office of Science and Technology Policy, U.S. Trade Representative and White House Legal Counsel. Federal agencies participating included: The U.S. Departments of Agriculture, Commerce (National Oceanic and Atmospheric Administration), Energy, Interior, Justice, State, Transportation, and Treasury. Other independent agencies included were the U.S. Environmental Protection Agency, National Aeronautics and Space Administration, and National Science Foundation. The report stated that "The best design for a climate change convention, and for any policy responses that might ensue, would be a 'comprehensive' approach that addresses all relevant trace gases, their sources and sinks...in order to deal with the many scientific, environmental and economic aspects of the climate system, which involves multiple trace gases resulting from activities in every sector of human society."

U.S. Congress, Office of Technology Assessment (OTA)

Changing by Degrees: Steps to Reduce Greenhouse Gases. U.S. Congress, Office of Technology Assessment, Washington, DC: February 1991. Report No. OTA-O-482. This assessment focuses principally on ways to cut carbon dioxide emissions in the United States and in other countries, although it does consider controls on other greenhouse gases. It states, "Major reductions of carbon dioxide and other greenhouse gases will require significant new initiatives by the federal government, the private sector, and individual citizens." The report considered programmatic requirements to reduce U.S. emissions by 15% by the year 2010. Authors concluded that, "Many of these initiatives will pay for themselves; for others, the economic cost may be considerable, [and that] many of these efforts need to be sustained over decades."

Preparing for an Uncertain Climate. U.S. Congress, Office of Technology Assessment, Washington, DC: October 1993. Report no. OTA-O-567. OTA's second report on climate change. This report was requested by three congressional committees: The Senate Committees on Environment and Public Works and on Commerce, Science and Transportation; and the House Committee on Science, Space, and Technology. The report, published in two volumes, identified more than 100 options that could help ease the transition to an uncertain climate, known as near-term, "targets of opportunity." This assessment addressed how natural and human systems may be affected by climate change; evaluated the tools at hand to ease adaptation to a warmer climate, considering coastal areas, water resources, agriculture, wetlands, preserved lands, and forests.

GLOBAL CLIMATE CHANGE SCIENCE OUTLOOK

The U.S. Senate has constitutional responsibility to provide advice and consent to ratification of international treaties which the United States has signed. Also, Congress has appropriations and oversight responsibilities for funding global change research activities and environmental research and development. Additionally, some individual lawmakers have called for enhanced research and development funding to devise fossil fuel emissions control technologies, lesser polluting technologies, energy conservation, and expanded use of nuclear power in the United States as possible voluntary solutions to reducing greenhouse gases. These are highly technical and complex issues that require expert scientific advice.

A number of hearings in the 106th Congress have addressed federal funding for climate change research, scientific debate about theoretical versus observed climate change, and the assumptions and findings of a variety of economic analyses that estimate the potential costs of U.S. implementation of the Kyoto Protocol, and the potential physical impacts of climate change. International efforts at negotiating climate change protection measures, in which the United States is a party, are continuing under the U.N. Framework Convention on Climate Change (FCCC), and negotiations related to the 1997 Kyoto Protocol.

In the spring of 1999, the U.S. Global Change Research Program (USGCRP) submitted performance measures for the first time under the 1993 Government Performance and Results Act (GPRA), for funds that were authorized in FY1998. Congress reviewed this information, to justify and prioritize research funding for global climate change research in the FY2000 budget. Congress also directed the White House to report on expenditures for domestic and international climate change programs for FY1998 and FY1999. These were transmitted to several congressional committees, pursuant to language in Title V of P.L. 105-277, The Foreign Operations, Export Finance, and Related Programs Appropriation Bill, FY1999.[11]

Late in the year 2000, the IPCC is expected to release a third scientific assessment on global climate change. New scientific findings and conclusions of the IPCC are likely to be revealed during expert review that might have bearing on U.S. policy and future negotiations on the Kyoto Protocol, to the extent that these occur. In addition, the USGCRP has prepared a National Assessment of the potential regional consequences of climate change for the United States, for which an overview, or synthesis report, is expected to be released in April 2000. Scientific knowledge about, and new research findings on, climate change will continue to play a role in the policy debate on key issues related to climate change concerns.

---

[11] See CRS Report 98-665, *Global Climate Change: Congressional Concern About "Back Door" Implementation of the 1997 U.N. Kyoto Protocol.*

Table 1. Major Conferences on Global Climate Change at which the U.S. Government has had Diplomatic Representation[1]

PERIOD	WHEN	WHAT	WHERE
Pre-UNCED	June 1998	The Changing Atmosphere (UNEP)	Toronto, Canada
	May 1989	Forum on Global Change	Washington, DC
	July 1989	Summit of the Arch (G-7)	Paris, France
	Nov. 1989	Ministerial Conference	Noordwijk, Neth.
	May 1990	Inter-parliamentary Conference	Washington, DC
	July 199	Economic Summit (G-7)	Houston, TX
	Nov. 1990	Second World Climate Conference	Geneva, Switzerland
	1989-1992	INC negotiations: U.N. FCCC adopted	New York, NY
UNCED (FCCC opened for signature)	June 1992	UNCED (Earth Summit)	Rio de Janeiro, Brazil
	March 1994	FCCC enters into force	
Post-UNCED	April 1995	COP-1, Berlin Mandate	Berlin, Germany
	July 1996	COP-2, Ministerial Declaration	Geneva, Switzerland
Pre-Kyoto	Oct. 1997	"Challenge of Global Warming" Conf.	Washington, DC
Kyoto	Dec. 1997	COP-3, U.N. Kyoto Protocol text adopted	Kyoto, Japan

[1] While this table is not exhaustive, it represents those meetings which many believe were significant milestones for United States participation in negotiations on international global climate protection agreements.

APPENDIX 1: A CHRONOLOGY OF U.S. GOVERNMENT INVOLVEMENT IN GLOBAL CLIMATE CHANGE POLICY THROUGH 1998

For decades, scientists in federal agencies, such as the National Aeronautics and Space Administration (NASA), National Oceanic and Atmospheric Administration (NOA), Environmental Protection Agency (EPA), and others, have participated in scientific workshops and international conferences on the nature of Earth's climate system, and the role of CO_2 and other greenhouse gases that are believed by many to modify the global climate. Extensive involvement of the United States government in formulating U.S. policy and assuming a diplomatic role in international efforts which relate to that issue, however, probably began around 1978, with efforts to coordinate federal government activities. The following chronology lists what are considered by many to be major events which shaped U.S. policy over the past 22 years.

1978 – The National Climate Program Act, P.L. 95-367, 15 USC § 2901, *et seq.*, established the National Climate Program Office (NCPO) in the National Oceanic and Atmospheric Administration (NOAA) of the Department of Commerce for the purposes of planning and coordinating U.S. involvement in international research efforts on climate change throughout the federal government.

February 1979 – The World Meteorological Organization (WMO), U.N. Environment Programme (UNEP), and International Council of Scientific Unions (ICSU) sponsored the First World Climate Conference (FWCC) in Geneva, Switzerland. Billed as "a conference of experts on climate and mankind," and focusing on the scientific basis of climatic change, the FWCC addressed issues of northern hemisphere cooking; severe winters that were occurring in the mid-latitudes of the United States and Central Europe; widespread drought and desertification in Sub-Sahara Africa, and public concern about famine and death resulting from the observed effects of climate change on some world agricultural systems. The U.S. government sent representatives to the FWCC; however, those were mostly expert scientists employed at U.S. scientific mission agencies. Government scientists attending such conferences participated in their capacity as scientists, not as representatives of their respective governments.

Late 1979 – Out of the FWCC evolved the WMO World Climate Program (WCP) jointly sponsored by WMO/UNEP/ICSU, and the four components of WCP: (1) the World Climate Data Program; (2) the World Climate Applications Program; (3) the World Climate Impact Studies Program; and (4) the World Climate Research Program. Each was dedicated to examining particular aspects of the state of scientific knowledge about climate change while deducing the technological capability of various nations to address global climate change. In a series of conferences and workshops sponsored by the WMO, UNEP, and ICSU, the seeds of interest were sown among governments for officially participating in such activities.

1980 – The first WCP joint WMO/UNEP/ICSU meeting of experts on the Assessment of the Role of CO_2 on Climate Variations and Their Impact was held in Villach, Austria in November. This meeting investigated how increasing greenhouse gas concentrations in the atmosphere could affect various regions of the Earth in the 21st Century. Participants also discussed the technical, financial, and institutional options for limiting or adapting to climatic changes.

1982 – The three representative organizations of the WCP (WMO, UNEP, and ICSU) met in October, in Geneva, Switzerland and recommended that continuing assessments of CO_2, believed to be responsible for global warming, be held every 5 years, starting from the first meeting in 1980. Following that meeting, an Interim Assessment was prepared.

1985 – A second WCP scientific conference was held in Villach, Austria, in October to follow up and update as assessment, originally prepared in 1980, of the role of increased CO_2 and other radiatively active greenhouse gases in climate variation and their associated impacts. Participants at this meeting concluded in a conference statement that, "As a result of the increasing concentrations of greenhouse gases, it is now believed that in the first half of the next century a rise of global mean temperature could occur which is greater than any in man's history." It was at this session that full-scale national government interaction with scientists took root because the WCP recommended policy actions to stem potential impacts of climate change from increasing concentrations of CO_2 and other greenhouse gases.

September 1987 – The United States completed international negotiations under the auspices of the WMO/UNEP toward an international treaty and regulatory annexes to protect the stratosphere from ozone depletion suspected to result from man-made chlorofluorocarbons (CFCs). These negotiations, and the guidelines for their conduct were set forth as early as 1985, in accordance with the Vienna Convention of the Prevention on Stratospheric Ozone Loss which the United States had previously ratified. By 1987, the Parties to the Vienna Convention had concluded negotiations on an international regulatory instrument and had opened for signature the 1987 Montreal Protocol on Substances that Deplete the Ozone Layer. Some public policy experts have credited the generally positive experiences of scientists and policymakers working together on ozone protection negotiations as facilitating the organization and the conduct of both the activities of the IPCC working groups, and the subsequent U.N. negotiations undertaken for the Framework Convention on Climate Change.

October 1987 – Two WCP workshops took place in Villach, Austria, and Bellagio, Italy, which led to the discussion of developing international policies for responding to climatic change. The justification was built upon the results of both the 1980 and 1985 WCP scientific assessments on CO_2. The WCP Advisory Group on Greenhouse Gases (AGGG) saw this meeting as an important step in policy development in response to possible climate change at the international level and, as such, a realization of a goal that was called for originally by the Villach conference in 1985. About 6 months after the Bellagio meeting, the governing bodies of WMO and UNEP requested the U.N. to establish an Intergovernmental Panel on Climate Change (IPCC) to address the issue of climate change, its environmental, economic and social impacts, and possible national and international responses to such changes, and invited nations to have full ministerial representation in future proceedings.

December 1987 – The White House Committee on Earth Sciences (CES) 2as established under the White House Office of Science and Technology Policy (OSTP)

Federal Coordinating Committee on Science, Technology and Engineering (FCCSET). The committee came to be called the Committee on Earth and Environmental Sciences (CEES), and was charged with the development of a 10-year U.S. Global Change Research Program, established by the National Global Change Research Act of 1990 (P.L. 101-606, 15 USC § 2921 *et seq.*).

June 1998 – The [U.N.] Toronto Conference on the Changing Atmosphere was held in June. Governments were invited by Canada to participate in formal discussions leading toward a possible "law of the atmosphere," controlling atmospheric pollutants which, among other things, would seek to control emissions of CO_2 into the atmosphere.

November 1988 – The first meeting of a newly created U.N. Intergovernmental Panel on Climate Change (IPCC) convened in Geneva, November 9-11, 1988. The plenary meeting of the IPCC included representative of 35 nations, including the United States, the U.S.S.R., several other foreign governments, and international governmental and non-governmental organizations. The latter served as observers and functioned as advisory bodies in the proceedings. As a result of this meeting, the IPCC was charged by the U.N. General Assembly to prepare an integrated state-of-the-art report on the science, impacts, and responses to global climate change by September 1990.

February 1989 – Following from the 1988 Toronto Conference, at a "Meeting of Legal and Policy Experts" held in Ottawa, on February 20-22, 1989, participants discussed the feasibility of a climate change convention and issued a statement that, among other things, addressed considerations and elements for a specific convention on climate change which would govern emissions of carbon dioxide implicated in global warming, and target a 20% reduction thereof.

Spring 1990 – Three consensus documents on science, impacts, and response of the IPCC working groups were produced by the beginning of summer of 1990 and were viewed throughout most of the international scientific and global diplomatic community as definitive statements of state-of-knowledge about global climate change. A majority of participants and independently polled scientists who peer reviewed those reports considered the results of each working group a relative success and a major accomplishment for multi-disciplinary scientific and social research bodies participating within a potential policy making milieu. The "responses" working group report was criticized for not offering concrete recommendations as to what governments should do to either mitigate or adapt to potential climate change. Furthermore, a group of dissenting scientists claimed that contrary opinions were neither considered nor presented in the final IPCC documents and, consequently, criticized the IPCC review process because no comments or reactions to comments were ever exchanged between independent peer reviewers and the IPCC.

June through August 1990 – The three IPCC working groups submitted their findings to the full IPCC in June 1990, and, following a plenary session in August 1990, the IPCC presented it *First Interim Assessment Report* to the 45th session of the U.N. General Assembly, at a session of U.N. Second World Climate Conference (SWCC) in Geneva, Switzerland between October 29 – November 7, 1990. SWCC was convened by WMO, UNEP, and ICSU. The IPCC's integrated "synthesis report" was adopted by the U.N. General Assembly and would form the basis for international negotiations toward a framework convention on climate change. The synthesis report consisted of the Interim Assessment, a fourth working paper prepared by the ad hoc IPCC Working Group on Financial and Technical Assistance, and an IPCC Assessment Overview prepared by the Secretariat of the IPCC. It was subject to a short period of review during SWCC Scientific and Technical sessions.

November 1990 – After presentation of the IPCC integrated Interim Assessment of Global Climate Change at the SWCC, and its adoption by the U.N. General Assembly, some countries had expected that negotiating sessions for a regulatory mechanism to address potential global climate change would begin during Ministerial sessions that immediately followed the Scientific and Technical sessions. The United States and the former Soviet Union, however, opposed making any commitments at that time, especially any legally binding reductions of CO_2 or other greenhouse gases. The United States argued that such matters would be more appropriately considered under the authority of the U.N. General Assembly and not the WMO, and further suggested that interested nations reconvene in negotiations that would address specific regulatory actions relating to global climate change.

December 1990 – The U.N. General Assembly on December 21, 1990, in furthering its resolutions 43/53 of December 6, 1988 and 44/207 of December 22, 1989, which acknowledged that climate change is a common concern of mankind, established an Intergovernmental Negotiating Committee (INC). The INC, supported by WMO and UNEP, was charged with preparing the future FCCC, which would contain appropriate commitments and any related legal instruments as might be agreed upon. This resolution, A/RES/45/212, called for the framework convention negotiations to be completed prior to the U.N. Conference on Environment and Development (UNCED) scheduled for the June 1992 "Earth Summit" in Rio de Janeiro, and to be opened for signature during that conference.

February 1991 – The United States offered to host the first INC session in Chantilly, VA, in February 1991. The meeting was titled, "Protection of global climate for present and future generations of mankind." Some criticized it as unproductive because no protocols, memoranda or understanding, or terms of reference relating to a framework convention on climate change came out of this first session. Others (such as the United States delegation) insisted that INC's focus at its first session was primarily to attend to organizational business and the INC's administrative requirements. By the close of deliberations, two subsidiary bodies on science and technology and implementation (and their leadership) had been established.

June 1991 through May 1992 – The second session of the INC met in Geneva, June 19-29, 1991; the third session convened in Nairobi, September 9-20; the fourth session convened in Geneva, December 9-20; and the fifth session convened in New York City, February 18-28, 1992. One more negotiating session, described as an extension of the 5th INC session, took place in New York between April 29 and May 8, 1992. This session was the last remaining opportunity for the parties to meet as a whole and agree upon a final text for a so-called framework convention on climate change that would be opened for signature in June at UNCED, in Rio de Janeiro. At the conclusion of this last session, it was evident that a flexible, voluntary response by nations to reduce net atmospheric concentrations of greenhouse gases would be the backbone of the climate convention. This agreement contained a "non-binding aim" of voluntary commitments for industrialized countries to begin to return their net emissions of greenhouse gases to 1990 levels and to devise plans for stabilizing concentrations of greenhouse gases in the atmosphere by 2000, both by controlling sources of emissions and enhancing sinks for greenhouse gases.

January 1992 – Resulting from a meeting in January in Guangzhou, China, the IPCC's Working Group-1 on Science released a "Supplement" to update the first interim scientific assessment of climate change. An IPCC plenary document was also released that integrated findings from activities of the other IPCC working groups. New scientific insights into the role of CFCs and climate change, as potentially offsetting some global warming at Earth's surface, challenged the Bush Administration's "basket of options" approach to reducing greenhouse gas emissions, and shifted emphasis in international negotiations back to focusing on CO_2 reductions. The IPCC also declared a need to reassess the indirect global warming potential of other greenhouse gases and their concentration and effect over different time horizons; and the IPCC also called for further study on the possible climatic cooling effects of sulfate aerosols in Earth's atmosphere.

February 1992 – In its *Statement on Commitments*, submitted at the 5th INC session, the United States outlined a new course of measures that it would undertake to mitigate climate change. The United States emphasized that these actions would begin immediately, would be taken unilaterally, and would not be contingent on its final acceptance or rejection of any legally binding timetables or provisions as might be set forth in a future international climate change agreement. Actions would be pursued in several areas, including: (1) improved energy efficiency; (2) transportation sector improvements; (3) supply-side changes to lower-emission technologies; (4) agriculture and natural resources 0 methane capture and tree planting; (5) federal research and development measures – technological and scientific; (6) joint U.S. government-industry programs to reduce emissions; and (7) state and local government actions. Some analysts estimated that such actions could reduce CO_2 by about 14% below 1990 levels by 2000. Environmentalists criticized these "new" measures as simply a delineation of what the United States had been prepared to do all along and, in some cases, what may have been required under existing law dealing with energy conservation.

April 1992 – A study by the Bush Administration, *U.S. Views on Climate Change*, suggested that the United States might not be far from the goal of reducing its net emissions of CO_2 to 1990 levels by 2000 – a goal called for by many INC parties – simply by undertaking energy efficiency and savings programs and other mitigation and adaptation strategies for climate change that were already underway in existing federal and state programs.

May 1992 – INC negotiations, which had begun in February 1991, concluded at the U.N. in New York City. A ministerial draft text was adopted culminating in an international treaty formally known as the U.N. Framework Convention on Climate Change.

June 1992 – On June 12, at UNCED (the Earth Summit) in Rio de Janeiro, the United States and 142 other nations signed the U.N. Framework Convention on Climate

Change (FCCC). The FCCC contained an action framework that would commit the world's industrialized countries to voluntary reduction of greenhouse gases and enhancing greenhouse gas sinks. Such actions would be aimed at stabilizing atmospheric concentrations of greenhouse gases at 1990 levels. The FCCC also contained other commitments for all signatory parties, including developing countries, related to its establishment, support, and administration. Furthermore, the Convention suggested the possibility of continuing negotiations through a Conference of Parties, subject to a judgment of the ratifying parties, after the FCCC's entry into force, that would meet periodically to pursue subsequent actions to counter global warming – similar to the 1985 Vienna Convention, that preceded the Montreal Protocol on Protection of the Ozone Layer.

The consensus view among INC representatives was that the convention opened for signature at UNCED represented a scientifically sound first step towards stabilizing industrial greenhouse gas emissions. Critics, however, found it deficient because in their view it did not realistically address greenhouse gas emissions from the transportation sector and emissions reductions were voluntary, rather than legally binding. Furthermore, there was disagreement on how far the convention should have gone, and whether it should have also set future emission reduction targets and timetables beyond the year 2000 target.

September through October 1992 – On September 8, 1992, the Convention was transmitted by the White House to the Senate Committee on Foreign Relations for the advice and consent of the Senate to ratification. That Committee endorsed the treaty and reported it (S. Exec. Rept. 102-55) on October 1. The Senate consented to ratification of the U.N. Framework Convention on Climate Change on October 7, 1992, with a two-thirds majority vote; President Bush signed the instrument of ratification of the FCCC on October 13, 1992.

Late November 1992 – The *U.S. National Action Plan for Global Climate Change* was released at the end of November by the Bush Administration to supplement many of the energy conservation measures in force that may have had a secondary effect of limiting future U.S. net greenhouse gas emissions. The former was known as a "no-regrets" strategy, because it made sense economically, regardless of any potential global warming effects. The National Action Plan also considered other efforts that might be undertaken to adapt to potential climate change, reiterating many of the strategies outlined in President Bush's earlier 1991 *Action Agenda on Climate Change*. However, the 1992 U.S. National Plan went further than the *Action Agenda* to include: (1) additional federal government measures, both legislative and administrative; (2) actions taken by state governments; (3) private sector measures; and (4) measures undertaken in cooperation with other countries.

December 1992 – The INC convened its sixth meeting December 7-10, in Geneva, Switzerland, to discuss its future, and to re-evaluate the urgency (timetable of meetings) of moving forward on measures to address potential global climate change. The INC was requested to act as the interim coordinating body on business matters relating to global climate change for the U.N. Secretary General, until the Conference of the Parties to the FCCC would be established and meet for the first time. The INC met in March and

August 1993, and, among other things, debated the feasibility of the World Bank's Global Environmental Fund (GEF) as a mechanism for managing international funding of developing countries, which would assist the latter in fulfilling their commitments and obligations under the FCCC.

October 1993 – President Clinton and Vice President Gore released a U.S. Climate Change Action Plan that suggested some 52 voluntary measures to reduce U.S. greenhouse gas emissions to 1990 levels by 2000. This domestic goal would be aligned, in principle, with U.S. commitments under the FCCC.

March 1994 – According to terms of the FCCC, the U.N. Secretary General having received at least 50 countries instruments of ratification, the Convention entered into force March 24, 1994.

April 1994 – Seeking grounds for a uniform approach toward climate protection, the first Conference of Parties (COP-1) to FCCC met in Berlin, Germany, in the spring of 1995, and voiced concerns about the adequacy of countries' abilities to meet commitments under the Convention. These were expressed in a U.N. ministerial declaration known as the "Berlin Mandate," which established a 2-year Analytical and Assessment Phase (AAP) to negotiate a "comprehensive menu of actions_ from which countries would choose options to address climate change that for them, individually, made the best economic and environmental sense. FCCC parties also deliberated over elements of possible amendments to FCCC and/or a subsequent protocol that might advance climate protection. They also discussed whether numerical aims, such as targets and timetables, binding or non-binding agreements, or technology-related goals, alone, might be "adequate" for climate protection.

Another major issue dealt with was what some called "an arbitrary division between Annex I and Developing countries" that concerned the effectiveness of commitments of each class of countries in achieving the goals of FCCC. Criticism was leveled by many industrialized countries, including the United States, against many newly industrializing countries (NICs), such as Brazil, India, and China, because NICs would continue to be classified as non-Annex I countries and enjoy certain exemptions under the Berlin Mandate – including exemptions from possible future, legally binding emissions reduction agreements – even though these countries collectively could become the world's largest emitters of greenhouse gas emissions with 15 years.[1]

July 1996 – The Second Conference of Parties to the FCCC (COP-2) met in July 1996 in Geneva, Switzerland, and its Ministerial Declaration was adopted July 18, 1996. This document reflected a U.S. position statement presented by Timothy Wirth (former Under Secretary for Global Affairs for the U.S. State Department) that: 1) accepted outright the scientific findings on climate change proffered by the IPCC in its second assessment (1995); 2) rejected uniform "harmonized policies" in favor of flexibility; and 3) called for "legally binding mid-term targets." Legally, the Declaration represented the consensus of ministerial participants at COP-2 that, as a body, they did not object to a

---

[1] For more information, see CRS Report 96-699, *Global Climate Change: Adequacy of Commitments Under the U.N. Framework convention and the Berlin Mandate.*

"future decision which would be binding on all parties under the FCCC," with individual reservations included and noted.

June 1997 – On June 26, 1997, President Clinton submitted "Additional U.S. Proposals" aimed at efforts to educate the American public on the need for a climate protection protocol and stronger climate protection measures. This plan included installing a million solar panels on roofs across the United States by the year 2010, convening a White House Conference (October 6, 1997); $1 billion in foreign aid for "best environmental practices," and an environmental technology R&D and trade incentive for U.S. industry (known as the Climate Change Technology Initiative). The vehicle to achieve these reductions was a 3-track proposal which included $5 billion in tax breaks (over 5 years) to U.S. industries to develop technologies and practices that reduce greenhouse gas emissions, a restructuring of the electric utilities industry, and the development of some form of emissions trading among FCCC parties for credits.

In addition, the State Department submitted "Additional U.S. Proposals" to the COP, that would: 1) condone penalties for parties who exceed an allowed emissions budget for a given 5-year period; 2) clarify eligibility of parties to participate in emissions trading schemes and their obligations pertaining to measurements and reporting of emissions, including devising national mechanisms for certification and verification of trades; and 3) preclude from trading any party which exceeds its emission budget or is in question of compliance. These proposals concerning treaty compliance were not adopted in the final working text of the Protocol ("the Kyoto Accord"), but are currently part of ongoing negotiations leading up to COP-6.

December 1997 – Prior to ministerial negotiations and COP-3, in Kyoto in December 1997, President Clinton had proposed the goal to return U.S. greenhouse gas emissions to 1990 levels by 2012 (a 30% reduction from estimated 2012 levels). Japan, on the other hand, had proposed a 5% reduction in CO_2 below 1990 levels by the year 2012, and the EU had proposed reducing emissions of three greenhouse gases by 7.5% by 2005, and then by 15% by 2010. The developing countries (G-77) and the Association of Small Island States (AOSIS) sought a reduction of CO_2 emissions 35% below 1990 levels, to apply to the industrialized countries, exempting themselves from emissions reductions.

The work on the U.N. Kyoto Protocol on Climate change was completed December 11, in Kyoto, Japan, in a half-day extension of the official session. Most industrialized nations and some central European countries (defined in "Annex B" to the Protocol) agreed to legally binding reductions in greenhouse gas emissions of an average of 55 below 1990 levels between the years 2008-2012, identified as the first emissions budget period. The United States would be required to reduce its total emissions an average of 7% below 1990 levels by the year 2012. However, some Annex-B countries (Australia, e.g.) would be allowed to increase their greenhouse gas emissions. Globally, emissions of three major greenhouse gases (CO_2, CH_4, N_2O) would be targeted to decline about 5% below 1990 levels over the next 10 years.

On November 12, 1998, President Clinton instructed a representative to sign the Kyoto Protocol to "lock-in" commitments he judged to be in the U.S. national interest that were achieved during negotiations. Subsequently, the United States became the 60th country to sign the treaty. This act drew protest by some in Congress because the Kyoto Protocol had not yet been debated by the U.S. Senate and many claimed that it was in violation of S.Res. 98, the Byrd/Hagel Resolution, introduced in July 1997, which required an economic analysis and legally binding emission reductions for all FCCC parties. Many in congress believe the Protocol would pose an unfair burden on industrialized countries, while exempting developing countries from any regulatory requirements. After signing the Kyoto Protocol, the President announced he would continue to pursue efforts to gain "meaningful" commitments from key developing countries over the next couple of years, before he would consider sending the treaty to the U.S. Senate in deference to S.Res. 98.

In the meantime, Congress has held hearings on the potential economic impacts of ratifying the Kyoto Protocol. Some Administration officials and supporters claim there are economic benefits to be realized from U.S. ratification of the treaty, while many opponents, including a number in Congress, have projected a significant negative impact on the U.S. Economy.

Chapter 4

CLIMATE CHANGE TECHNOLOGY INITIATIVE (CCTI): RESEARCH, TECHNOLOGY, AND RELATED PROGRAMS

Michael M. Simpson

BACKGROUND

The Climate Change Technology Initiative was described as "the cornerstone of the (Clinton) Administration's efforts to stimulate the development and use of renewable energy technologies and energy efficiency products that will help reduce greenhouse gas emissions,"[1] through a combination of research and development (R&D), and information and tax incentive programs. Carbon dioxide, the major "greenhouse gas" of concern in possible climate change, is produced in large part as a result of energy production and use when these are based on fossil fuel combustion. The federal government has had programs dealing with energy efficiency for more than 20 years, and the Congress has held hearings about them since the mid-1970s, when a major goal of such programs was to reduce U.S. dependence on oil imports during the energy crisis.

U.S. government policies explicitly addressing possible climate change linked to "greenhouse gas" emissions date back to the mid-1980s.[2] These policies have focused heavily on scientific research. The Energy Policy Act of 1992, in conjunction with the U.S. ratification of the 1992 United Nations Framework Convention on Climate Change (UNFCCC), set the direction of U.S. efforts under the Bush and Clinton Administrations toward energy efficiency, renewable energy, and R&D[3], to try to move toward stabilizing

---

[1] Testimony on May 20, 1999 by Deidre A. Lee, Acting Deputy Director for Management, Office of Management and Budget (OMB), to the House Committee on Government Reform and Oversight, Subcommittee on National Economic Growth.

[2] For details, please see CRS Issue Brief IB89005, *Global Climate Change*.

[3] For further details on this, please see CRS Report RL30024, *"Global Climate Change Policy" From "No Regrets" to S.Res. 98*.

atmospheric greenhouse gas concentrations.[4] The Climate Change Action Plan announced in 1993 included more than 40 federal programs working with business, state and local governments, and other entities with the goal of reducing U.S. greenhouse gas emissions. R&D and other programs since then had largely been maintained or extended, or modified with some new activities and names. With evolution from the hybridization among prior efforts, couple with some augmentation, packages of programs in the Clinton Administration such as the CCTI were build upon these earlier efforts, including efforts to reduce dependence on oil imports.

During the preparations for the final negotiations of the December 1997 Kyoto Protocol to the UNFCCC,[5] President Clinton announced a three-stage climate change plan on October 22, 1997.[6] Stage 1, as announced in 1997, including funding for research and development (R&D), tax incentives for early action, a set of federal government energy initiatives including various tax credits to encourage purchase and use of more efficient technologies, and industry consultations to explore ways to reduce greenhouse gas emissions. Stage 2, expected to begin around 2004, would review and evaluate stage 1. Stage 3, as envisioned prior to Kyoto, included actions aimed at reducing emissions to 1990 levels by 2008-2012, meeting the binding targets the U.S. expected to be in the Kyoto Protocol through measures that include domestic and international emissions trading. The Kyoto Protocol (which the United States signed on November 11, 1998 but which has not been submitted to the U.S. Senate for advice and consent on ratification) outlines an obligation for the United States to reduce its total greenhouse gas emissions by an average of 7% below 1990 levels between 2008 and 2012.[7]

The Congress has passed budget resolutions and appropriations bills with provisions prohibiting the use of funds to implement the Kyoto Protocol, which has not been ratified by the United States or entered into force internationally. Some controversy has been engendered by the possible linkage of funding proposals associated with the CCTI to the Kyoto Protocol goals. After some early consideration of these concerns, for the most part the R&D elements have been acceptable to the Congress. Moreover, many of the programs related to the CCTI and other climate research preceded the Kyoto Protocol, and in fact would be relevant to the voluntary commitments the United States has made in the U.N. Framework Convention on Climate Change to try to meet a voluntary goal of returning greenhouse gas emissions to 1990 levels.

As first outlined in President Clinton's FY1999 budget request[8], the CCTI was to be a combination of research and technology programs and of tax incentives to accelerate development and deployment of technologies designed to reduce greenhouse gas emissions. "The CCTI builds and expands upon an existing foundation of advanced science, basic research, and government-industry partnership. It will increase U.S.

---

[4] One example is the Climate Change Action Plan, released by President Clinton on October 19, 1993, which proposed voluntary domestic measures for stabilizing greenhouse gas emissions.

[5] Please see CRS Report 98-2 *Global Climate Change Treaty: The Kyoto Protocol* for details.

[6] Details about the plan, as set forth in 1997, can be found at [http://www.whitehouse.gov/Initiatives/Climate/3stage.html].

[7] Please see CRS Report 98-2 *Global Climate Change Treaty: Summary of the Kyoto Protocol* for further details.

[8] See [http://www.usia.gov/topical/global/environ/climate/whfs1198/ccti.html] for details.

competitiveness, reduce U.S. dependence on foreign oil, help maintain U.S. leadership in energy technology, and reduce greenhouse gas emissions at the same time."[9]

FEDERAL FUNDING LEVELS

CCTI funding consisted of two basic parts: (1) research and technology programs, and (2) targeted tax incentives. The research and technology program in turn consisted of two main parts: research and development, which primarily focused on understanding processes and developing new technologies related to carbon sequestration and to energy efficiencies; and information, audit, and other assistance programs to facilitate diffusion of technologies designed to improve energy efficiency or otherwise diminish greenhouse gas emissions. These two main parts of the research and technology side of CCTI were not always clearly distinct; to some extent there was a continuum with R&D at one end and assistance programs at the other. Nonetheless, the distinction has proved significant, in that R&D was noncontroversial, while the assistance programs had been, as some argued that market forces should have been allowed to determine commercial development and application. (The same objections were lodged against the tax incentive proposals.)

As enacted for FY1999, $1.021 billion went to research and technology programs and no funds were provided for tax incentives. As described in subsequent Clinton Administration documents, President Clinton's climate change plans were enlarged beyond the CCTI to include a proposed Clean Air Partnership Fund to support government and private efforts to reduce greenhouse gas emissions and ground-level air pollutants, work toward legislation on possible credit to companies for early voluntary action to reduce greenhouse gas emissions or increase carbon sequestration, and continuation of diplomatic efforts to develop details in the Kyoto Protocol on such matters as international emissions trading and participation by developing countries. This report discusses only the research and technology activities (which were basic R&D and information programs), and related funding aspects of the CCTI.[10]

The FY2001 request for the research and technology element of CCTI was $1.432 billion. Also requested was $4.030 billion for a 5-year package of targeted tax incentives,[11] not covered in this report. As shown in Table 1, by far the largest portion of CCTI research and technology funding was to go to the Department of Energy (DOE: 89% of the FY2000 overall CCTI budget as enacted; 81% of the FY2001 request) and to the Environmental Protection Agency (EPA: 10% of the FY2000 overall CCTI budget as enacted; 16% of the FY2001 request), with relatively small amounts to the Housing and Urban Development Department (HUD), the U.S. Department of Agriculture (USDA), and the Department of Commerce. It should be noted that as enacted in FY2000, while

---

[9] Climate Change Technology Initiative. A White House Fact Sheet, November 1998.
[10] See White House website [http://www.whitehouse.gov/WH/SOTU99/climate.html] for a general description of the President's climate change plan.
[11] Charter for hearing on Fiscal Year 2001 Climate Change Budget Authorization Request, House Committee on Science, Subcommittee on Energy and Environment, March 9, 2000. Page 5.

DOE received 87% of its FY2000 request, EPA was given 48% of its request, reflecting concerns raised about non-R&D activities.

Historically, as part of the FY1999 Clinton Administration budget proposals, President Clinton in February 1998 first proposed the Climate Change Technology Initiative. It proposed funding primarily for research and development activities at the Department of Energy, tax credits to encourage purchases of certain energy-efficient cars and houses, EPA's voluntary information programs to encourage businesses and others to conserve energy, and research into ways to sequester carbon in agriculture, in some cases as renewable fuels. In general in the CCTI, R&D relating to energy efficiency and renewable energy sources were largely evolutionary steps from earlier programs, initiated in the late 1970s and early 1980s to reduce dependency on oil imports.

Speaking about the Clinton Administration's FY2000 CCTI budget requests, a senior DOE official said "although the tax credits are largely new initiatives, many of the other programs are continuations or expansions of ongoing research, development, and deployment programs."[12] The CCTI (composed of R&D, incentive, and voluntary information programs) grew from the base programs detailed in "The Climate Change Action Plan" (released by the U.S. in October 1993) with consultations among the Federal entities (including the Global Change Research Program) and the Office of Management and Budget.

Table 1. CCTI Research and Technology Funding by Agency ($ Millions)

Department/ Agency	FY1998 enacted	FY1999 enacted	FY2000 request	FY2000 enacted	FY2001 request; [% of CCTI FY2001 request
Department of Energy	729	902	1124	980	1169 [81]
Environmental Protection Agency	90	109	216	103	227 [16]
Housing and Urban Development	0	10	10	10	12 [1]
U.S. Department of Agriculture	0	0	16	0	24 [2]
Department of Commerce	0	0	2	2	0 [0]
TOTAL	819	1021	1368	1095	1432

Source: "President Clinton's FY2001 Climate Change Budget," page 13.

---

[12] Testimony on April 14, 1999 by Jay Hakes, Administrator, Energy Information Administration, U.S. Department of Energy, to the House Committee on Science.

Department of Energy

Carbon dioxide, the major greenhouse gas, rises mostly from combustion of fossil fuels. The Department of Energy, which has long had R&D programs relating to fossil fuel energy use from its days seeking to manage and to develop energy supplies, was by far the largest recipient of CCTI funding. DOE received $980 million for CCTI activities in FY2000 (98% of all federal CCTI funds), approximately 87% of the level of funding that is requested. DOE received from 82% to 89% of the total funding for the Initiative. Funding for the DOE's efforts in the CCTI were planned for the research, development, and deployment of more energy efficient and renewable technologies such as:

- For "buildings," low-power sulfur lamps, advanced heat pumps, chillers and commercial refrigeration, fuel cells, insulation, energy conserving building materials, and advanced windows;
- For "Electricity," generation using alternatives to fossil fuels such as solar energy, biomass power, wind energy, geothermal power, hydropower, and optimized nuclear power;
- For more efficient "Industries" including aluminum, steel, mining, agriculture, chemicals, forest products, and petroleum;
- For researching, developing, and deploying more efficient "Transportation" technologies, including furthering the Partnership for a New Generation of Vehicles (PNGV), a 10-year government/domestic auto industry partnership that aims to produce by 2004 a prototype midsize family car with 80 miles per gallon gasoline efficiency and a two-thirds reduction in carbon emissions; seven federal agencies are involved in the PNGV (Commerce, Defense, Energy, Transportation, EPA, National Aeronautics and Space Administration, and the National Science Foundation);
- For trying to find better ways to "Remove and Sequester Carbon" from fossil and other fuels, via agricultural and other approaches (in conjunction with EPA, and originally planned in conjunction with USDA); and
- For government efforts (federal, state, and others) to conserve energy through more highly coordinated "Management, Planning, Analysis and Outreach."[13]

As with the PNGV program, many of DOE's CCTI research and technology dollars were spent in partnership with other federal entities such as EPA and HUD, with other governmental units, and with private sector entities. As noted above, many of DOE's activities identified as Climate Change Technology Initiative, were to a great extent a continuation or evolution of DOE (and other federal) programs that predate the CCTI (and predate the 1977 establishment of the DOE in some cases, e.g., research into energy conservation and renewable energy sources). All of DOE's FY2000 CCTI funding and programs were continuations of FY1999 programs.

---

[13] Analysis of the Climate Change Technology Initiative, Research and Development Support. Energy Information Agency, U.S. Department of Energy. [http://www.eia.doe.gov/oiaf/climate99/research.html]

See Table 2 for a breakdown of funding levels for the DOE CCTI research and technology programs. Specific program and funding details for FY2000 and prior years were released in May 1999, as show in the Source not of Table 2.

Environmental Protection Agency

The Environmental Protection Agency uses two main budget categories: Science and Technology (S&T, which includes R&D and technology development and diffusion efforts), and Environmental Programs and Management (EPM, which are the costs to run programs). Therefore, it is difficult to consistently separate R&D from technology assistance and diffusion efforts. For example, in EPA's CCTI Buildings Sector, the owner of a building can have EPA's benchmarking tool voluntarily applied to that building as a target for energy use. Various activities can be tried, e.g., plugging leaks and replacing less efficient lights with more efficient lights, to see if the benchmark will be met. If not, other activities can be tried in an iterative fashion, trying and recording and incorporating the findings in the benchmark. This program includes activities that can be described as both research-related and technology diffusion and assistance. EPA's figures for CCTI S&T are used here.

The EPA in Fy2000 received $103 million for CCTI research and technology activities (about 9% of all the federal CCTI research and technology funds), a distant second to DOE's $980 million (89% of all federal CCTI research and technology funds). Also notable is the fact that while DOE received 87% of its FY2000 request, EPA got 48% of its request. While there has been some discussion about the proper roles for government, industry, and academe in climate change and other R&D,[14] the CCTI R&D activities were not highly controversial. In general, EPA funds targeted for R&D, especially areas of more basic R&D that predate the CCTI and the Kyoto Protocol, were less controversial, and funds for new programs intended to assist technology deployment and diffusion and to help consumers learn about and choose more efficient commodities and processes were more controversial.

The elements and levels of EPA's CCTI research and technology funds are summarized in the following table. Activities related to these program areas are briefly described below. Some of these finding areas focused heavily on R&D, while others involved information dissemination and other activities.

---

[14] Please see CRS Report 98-365 *Some Perspectives on the Changing Role of the U.S. Government in Science and Technology* for details.

Table 2. DOE CCTI Research and Technology Programs ($Millions)

Program	FY98 actual	FY99 estimate	FY00 proposed	FY00 enacted	FY01* request
Buildings	102	124	183	141	—
Energy Conservation	79	96	145	115	
Energy Conservation (Federal Buildings)	20	24	32	24	
Solar Ht/Cool/Hot Water	3	4	6	2	
Transportation	223	250	316	274	—
Energy Conservation	193	202	252	232	
Solar/Renewable, Alternative Fuels	30	42	53	39	
Energy Information Administration	—	3	3	3	
Basic Science	—	3	8	**	
Industry	136	167	172	170	—
Energy Conservation	136	166	171	170	
Basic Science	—	1	1	**	
Electricity	239	311	375	307	—
Solar/Renewable	239	291	340	268	
Nuclear	0	0	5	5	
Fossil	—	18	27	34	
Basic Science	—	2	3	**	
Carbon Removal & Sequestration	—	13	29	9	—
Fossil	—	6	9	9	
Basic Science	—	7	20	**	
Management, Planning, Analysis & Outreach	29	38	47	43	—
Energy Conservation	29	38	47	43	
Basic Science	**	**	**	33	—
Total (may not add due to rounding)	729	902	1124	976	1169

Source: U.S. Department of Energy. "Department of Energy Reports to Congress on FY2000 Expenditures for Energy Supply, Efficiency, and Security Technologies Supporting the Climate Change Technology Initiative" May 18, 1999. p. 3. "FY2000 Enacted" expenditures were obtained via telephone conservation on December 20, 1999 from the Department of Energy, and include estimates for the spread of the 0.38% rescission.

* Details were unavailable as of March 13, 2000.

** "Basic Science" was presented in FY2000 for the first time as a specific category. It had been funded before in a fragmented fashion throughout other categories.

Table 3. EPA CCTI Research and Technology Programs ($ Millions)

Program	FY99 request	FY99 enacted	FY00 request	FY00 enacted	FY01 request
Buildings	78.1	38.8	80.1	42.6	80.1
Transportation	58.9	31.8	62.0	29.6	65.1
Industry	51.6	18.6	55.6	22.0	63.7
Carbon Removal	3.4	0.0	3.4	1.0	3.4
State & Local Governments	5.0	2.9	5.0	2.5	4.5
International Capacity Building	8.4	7.4	10.4	5.6	10.6
Research	0	10*	0*	0	0
Total	**205.4**	**109.5**	**216.5**	**103.5**	**227.4**

Source: (for all but FY00 and 01) EPA FY2000 Annual Performance Plan and Congressional Justification, p. VI-19 and HR1743 "Environmental Protection Agency Office of Air and Radiation Authorization Act of 1999" ordered to be reported May 26, 1999.

*From the EPA FY2000 Annual Performance Plan, p. VI-33, "Funding is discontinued for Climate Change Technology Initiative activities funded through the FY1999 Omnibus appropriation." FY00 enacted and FY01 figures were obtained from EPA at [www.epa.gov/ocfo/budget/budget.htm] on February 8, 2000.

➢ The "Buildings" component of EPA's research and technology activities in the CCTI included housing and commercial structures. It had been argued by EPA and others (including DOE) that efforts by individual and organizational consumers to secure the most energy efficient process or commodity are hampered by a lack of objective information on which to make comparisons (for details, please see IB10020 *Energy Efficiency: Budget, Climate Change, and Electricity Restructuring Issues*). Through the Agency's ENERGY STAR Program and ENERGY STAR Buildings and Green Lights Partnership, EPA evaluates and certifies energy-saving building-related products (including such items as televisions, appliances, residential lighting, and whole houses), and makes that information available so that consumers and businesses can choose energy-saving and pollution-reducing products more easily.

➢ "Transportation" activities of EPA included the following:

- continued work in the Partnership for a New Generation of Vehicles (the government/domestic auto industry partnership described previously under DOE);
- expanded support for a program which provided new incentives for commuters to consider transit, ridesharing, or other alternatives to driving;
- continued support of state and local efforts toward livable communities and smart growth; and

- continued efforts in the Transportation Partners network which linked about 340 local governments, community organizations, and companies in order to produce knowledge that was designed to reduce vehicle miles traveled.

➤ EPA's "Industry" efforts included working with industries (especially energy-intensive industries such as cement, chemicals, steel, petroleum, airlines, and food processing), commonly through technical assistance, to audit and identify greenhouse gas emission sources and to help in formulating appropriate reduction goals and strategies, including removal of regulatory and other barriers. This included working with ongoing privately funded energy efficiency programs at primate companies.

➤ "Carbon Removal" efforts at EPA were planned in coordination with the Department of Agriculture. The EPA/USDA planned to use funds for this activity to study the kinds and sizes of incentives that could have been given to land owners and crop growers to increase the quantity of carbon stored on agricultural and forest lands, and at the same time improve soil quality, reduce soil erosion, and enhance other environmental and conservation goals;

➤ EPA worked with "State and Local Governments" to help find ways to reduce energy use and pollution, sometimes by supporting existing state and local programs. The Cities for Climate Protection program, for example, involved 54 local governments in 1998 to implement building, transportation, waste, and renewable energy projects to eliminate about 3 million metric tons of carbon dioxide. A state-level example is New Jersey's state carbon bank program, established to help achieve New Jersey's greenhouse gas emissions reduction goal of 3.5 % below 1990 levels by 2005.

➤ Developing countries currently emit more than half the global total of greenhouse gases, and such emissions are growing rapidly. "International Capacity Building" involved EPA and other agencies working to study ways to secure meaningful participation from key developing countries to reduce their emissions.

Department of Housing and Urban Development

CCTI research and technology programs were new to the Department of Housing and Urban Development (HUD) in FY1999, and the FY2000 budget proposed and received $10 million (no change from FY1999) for the government/housing developers/builders Partnership for Advancing Technology in Housing (PATH). Administered by HUD and identified as part of the CCTI, PATH research had a number of goals in addition to climate change. PATH efforts sought "to develop and disseminate technologies that will result in housing that is substantially more affordable, durable, disaster resistant, safer and energy/resource efficient..."[15] The FY2001 request was for $12 million.

---

[15] Department of Housing and Urban Development Policy Development and Research, from [http://www.hud.gov/bdfy2000/summary/pdandr/randt.html]

Department of Agriculture

The FY2000 request of $16 million for USDA's CCTI research and technology activities included $7 million for the Agricultural Research Service, $3 million for the Natural Resource Conservation Service, and $6 million for the Forest Service, to understand and better manage the carbon cycle, from sources to sequestration, focusing principally on agricultural approaches. While some of those proposed efforts were to build on prior work, the specific designation of USDA programs as part of CCTI funding was new in FY2000. However, the USDA identified a much wider array of base programs that carry out climate related research. The USDA overall had some $55 million on climate-related research among some 5 USDA agencies, an amount that had been stable during the 1990s.[16] The FY2000 appropriations as enacted contained no CCTI funds for USDA. The total FY2001 request of $24 million for CCTI activities was divided into $14 million for developing advanced biomass energy technologies, $6 million for studying agricultural carbon sequestration, and $4 million for examining agricultural practices and their relationships with greenhouse gas emissions.

Department of Commerce

Various base programs within the Department of Commerce addressed issues relating to climate change. The wide range of research in Commerce's National Oceanic and Atmospheric Administration (NOAA) included long-standing climate-related work, much of it not specifically identified as CCTI but rather part of NOAA's generic mission. Among other things, research at NOAA sought to determine "the chemical, and dynamical processes interact in the upper troposphere/lower stratosphere to affect climate; ...(and) study the effects of climate variability and change on health..."[17] There also were base programs at the National Institute of Standards and Technology (NIST) which looked at climate change issues.[18] The $2 million requested and provided in the FY2000 budget for the CCTI specifically was new to the Department and did not go to NOAA[19] or NIST as a single CCTI line item. No funds were specified for Commerce Department CCTI activities in the FY2001 budget request.

CONCLUSION

There were two parts to the research and technology elements of the CCTI: (1) R&D of environmentally more beneficial technologies and policies; and (2) information, audit, and other assistance intended to help individual and organizational consumers learn of,

---

[16] Telephone communication with the United States Department of Agriculture, Office of the Chief Economist, on September 13, 1999.

[17] Department of Commerce budget initiative, details of which can be found at [http://www.oarhq.noaa.gov/]

[18] Telephone communication with the National Institute of Standards and Technology on December 6, 1999.

[19] Personal communication with the National Oceanic and Atmospheric Administration on December 6, 1999.

choose, and use more efficient goods and processes (e.g., energy saving computers or industrial processes.

The pursuit of R&D was not highly controversial, especially for basic research. More controversy arose from the federal government's past and proposed efforts to use public funds to encourage and to help private individuals, companies, and organizations more quickly benefit from various environmental technologies. As state by then OMB Acting Deputy Director for Management Deidre Lee, spurring broader use of energy efficient technologies and renewable energy would have reduced energy bills and secure other benefits, so that "even if the threat of global warming did not exist, the (Clinton) Administration believes that these (CCTI) programs make good sense because they help our country address other energy-related and environmental challenges."[20] It was argued by some that economic benefits and saving money should have been sufficient incentives for consumers to invest in more efficient technology, that the renewable energy industry should have relied for commercial development on market forces rather than federal tax credits and information programs.[21] On the other hand, Lee and others argued that the Government needed to be involved to help overcome market barriers, such as a lack of accurate information, so as to permit informed energy-saving choices.

The CCTI was an effort by the Clinton Administration to draw on several federal agencies and departments in addressing the issue of climate change while securing other societal benefits as well. While the Clinton Administration's budget requests for CCTI R&D activities generated little controversy, its requests for CCTI information and tax incentive programs were more controversial. Differences between the Clinton Administration and Congress on the value of information and incentive programs in the various federal agencies and departments existed not only because of different perspectives between the executive and legislative branches, but also because of procedural and jurisdictional boundaries among the congressional committees and subcommittees responsible for the various federal agencies and departments. These boundaries made difficult tradeoffs among the several elements of CCTI and meant that each element tended to fend for itself in budgetary considerations.

---

[20] Testimony on May 20, 1999 by Deidre A. Lee, Acting Deputy Director for Management, OMB, to the House Committee on Government Reform and Oversight, Subcommittee on National Economic Growth.

[21] This position was described by Hon. Ken Calvert in his opening statement of the House Science Committee hearing on April 14, 1999.

Chapter 5

GLOBAL CLIMATE CHANGE: THE KYOTO PROTOCOL

Susan R. Fletcher

BACKGROUND

Responding to concerns that human activities are increasing concentrations of "greenhouse gases" (such as carbon dioxide and methane) in the atmosphere, most nations of the world joined together in 1992 to sign the United Nations Framework Convention on Climate Change (UNFCC). The United States was one of the first nations to ratify this treaty. It included a legally non-binding, voluntary pledge that the major industrialized/developed nations would reduce their greenhouse gas emissions to 1990 levels by the year 2000, and that all nations would undertake voluntary actions to measure, report, and limit greenhouse gas emissions.

However, as scientific consensus grew that human activities are having a discernible impact on global climate systems, possibly causing a warming of the Earth that could result in significant impacts such as sea level rise, changes in weather patterns and health effects – and as it became apparent that major nations such as the United States and Japan would not meet the voluntary stabilization target by 2000 – parties to the treaty decided in 1995 to enter into negotiations on a protocol to establish legally binding imitations or reductions in greenhouse gas emissions. It was decided by the Parties that this round of negotiations would establish limitations only for the developed countries (those listed in Annex I to the UNFCCC, including the former Communist countries, and referred to as "Annex I countries." Developing countries are referred to as "non-Annex I countries").[1]

During negotiations that preceded the December 1-11, 1997, meeting in Kyoto, Japan, little progress was made, and the most difficult issues were not resolved until the

---

[1] For additional information on the negotiations in Kyoto and related background, see CRS Report 97-1000, *Global Climate Change Treaty: Negotiations and Related Issues*; and CRS Issue Brief IB89005, *Global Climate Change.*

final days – and hours – of the Conference. There was wide disparity among key players especially on three items: (1) the amount of binding reductions in greenhouse gases to be required, and the gases to be included in these requirements; (2) whether developing countries should be part of the requirements for greenhouse gas limitations; and (3) whether to include emissions trading and joint implementation, (which allow credit to be given for emissions reductions to a country that provides funding or investments in other countries that bring about the actual reductions in those other countries or locations where they may be cheaper to attain).

Following completion of the Protocol in December of 1997, details of a number of the more difficult issues remained to be negotiated and resolved (see below). At the fourth Conference of the Parties (COP-4) held November 2-13, 1998, in Buenos Aires, Argentina, it was apparent that these issues could not be resolved as had been expected during this meeting. Instead, parties established a two-year "Buenos Aires Plan of Action" to deal with these issues, with a deadline for completion at the COP-6 meeting in The Hague, November 13-24, 2000. The difficulty in resolving these issues was underlined by the collapse of discussions in The Hague without agreement. A "COP-6 bis" will attempt to continue the negotiations, probably in May 2001 in Bonn, Germany.

MAJOR PROVISIONS OF THE KYOTO PROTOCOL

The Kyoto Protocol was opened for signature March 16, 1998, for one year, and would enter into force – become legally binding for countries that have ratified – when 55 nations have ratified it, provided that these ratification include Annex I Parties that account for at least 55% of total carbon dioxide emissions in 1990. this provision is likely to be hard to meet in the absence of U.S. ratification, since the United States accounted for well over 20 percent of such emissions. On November 12, 1998, the United States signed the Protocol, in part because the Clinton Administration wanted to revitalize what was seen as some loss of momentum during COP-4. As of November 27, 2000, 84 countries had signed the treaty, including the European Union and most of its members, Canada, Japan, China, and a range of developing countries. Some 31 countries were reported by the UNFCCC Secretariat to have ratified the treaty, but to date, no Annex I developed nation has ratified the Protocol. Nations are not subject to its commitments unless they have ratified it and it enters into force.

The major commitments in the treaty on the most controversial issues are as follows:

Emissions Reductions. The United States would be obligated under the Protocol to a cumulative reduction in its greenhouse gas emissions of 7% below 1990 levels for three major greenhouse gases, including carbon dioxide, (and below 1995 levels for the three other, man-mad gases), averaged over the commitment period 2008-2012. The Protocol states that Annex I Parties are committed – individually or jointly – to ensuring that their aggregate anthropogenic carbon dioxide equivalent emissions of greenhouse gases do not exceed amounts assigned to each country in Annex B to the Protocol, "with a view to reducing their overall emissions of such gases by at least 5% below 1990 levels in the commitment period 2008-2012." Annex A lists the 6 major greenhouse gases covered by the treaty.[2]

Annex B to the Kyoto Protocol lists 39 nations, including the United States, the European Union plus the individual EU nations, Japan, and many of the former Communist nations (the same countries as Annex I to the UNFCCC). The amounts for each country are listed as percentages of the base year, 1990 (except for some former Communist countries), and range from 92% (a reduction of 8%) for most European countries – to 110% (an increase of 10%) for Iceland. The United States agreed to a commitment on this list to 93%, or a reduction of 7%, to be achieved as an average over the 5-year commitment period, 2008-2012.

Based on projections of the growth of emissions using current technologies and processes, the reduction in greenhouse gas emissions required of the United States would likely be between 20% and 30% below where it would be otherwise by the 2008-2012 budget period.[3] However, inclusion of greenhouse gas sinks[4] - which the Protocol adopted as urged by the United States – and emission trading, means that the domestic U.S. emission reductions from fossil fuels needed to meet a 7% target would be substantially less. However, two of the most difficult issues unresolved at Kyoto, and responsible in large part for the breakdown of the COP-6 negotiations in November 2000, are related to (1) emissions trading – specifically, how much of a country's obligation to reduce emissions can be met through purchasing credits from outside, vs. taking domestic actions; and (2) the extent to which carbon sequestration by forests, soils, and agricultural practices can be counted toward a country's emission reductions.

Developing Country Responsibilities. The United States had taken a firm position that "meaningful participation" of developing countries in commitments made in the Protocol is critical both to achieving the goals of the treaty and to is approval by the U.S. Senate. This reflects the requirement articulated in S.Res.98, passed in mid-1997, that the United States should not become a party to the Kyoto Protocol until developing countries are subject to binding emissions targets. The U.S. government also argued that success in dealing with the issue of climate change and global warming would require such participation. The developing country bloc argued that the Berlin Mandate – the terms of reference of the Kyoto negotiations established at COP-1 in 1995 0 clearly excluded them from new commitments in this Protocol, and they continued to oppose emissions limitation commitments by non-Annex I countries.

The Kyoto negotiations concluded without such commitments, and the Clinton Administration indicated that it would not submit the Protocol for Senate consideration until meaningful commitments were made by developing countries. At the COP-4 in Buenos Aires, Argentine became the first nation to indicate that is will make a

---

[2] The six gases covered by the Protocol are carbon dioxide (CO_2), methane (CH_4), nitrous oxide (N_2O), hydro fluorocarbons (HFCs), per fluorocarbons (PFCs), and sulfur hexafluoride (SF_6). The most prominent of these, and the most pervasive in human economic activity is carbon dioxide, produced when wood or fossil fuels such as oil, coal, and gas are burned.

[3] See CRS Report 98-235 ENR, *Reducing Greenhouse Gases: How Much from What Baseline?*

[4] Greenhouse gases, especially CO_2, are absorbed by a number of processes in forests, soils, and other ecosystems. These are called "sinks."

commitment to take on a binding emissions target for the period 2008-2012. Kazakhstan also announced its intention to take similar action. At this meeting, the United States announced it would sign the Kyoto Protocol, which it did on November 12, 1998. To date, none of the largest developing countries, such as China, India, or Brazil, have shown a willingness to make commitments to reducing greenhouse gas emissions. The United States has continued to attempt raising this issue at COP meetings, generally under the agenda item "adequacy of commitments." However, at the COP-6 meeting, this item was not discussed, and developing country resistance continues.

The Protocol does call on all Parties – developed and developing – to take a number of steps to formulate national and regional programs to improve "local emission factors," activity data, models, and national inventories of greenhouse gas emissions and sinks that remove these gases from the atmosphere. All Parties are also committed to formulate, publish, and update climate change mitigation and adaptation measures, and to cooperate in promotion and transfer of environmentally sound technologies and in scientific and technical research on the climate system.

Emissions Trading and Joint Implementation. Emissions trading, in which a Party included in Annex I "may transfer to, or acquire from, any other such Party emission reduction units resulting from projects aimed at reducing anthropogenic emissions by sources or enhancing anthropogenic removals by sinks of greenhouse gases" for the purpose of meeting its commitments under the treaty, is allowed and outline in Article 6, with several provisos. Among the provisos is the requirement that such trading "shall be supplemental to domestic actions." The purpose of this proviso is to make it clear that a nation cannot entirely fulfill its responsibility to reduce domestic emissions by relying primarily on emissions trading or joint implementation to meet its targets. Joint implementation is project-based activity in which one country can receive emission reduction credits when it funds a project in another country where the emissions are actually reduced. One of the more contentious issues in on-going negotiations concerning how the Kyoto Protocol would work is this issue of "supplementarity" – finding agreement on what proportion of a nation's obligations could be met through these mechanisms versus domestic actions to reduce emissions within a nation's own borders.

A number of specific issues related to the rules on how joint implementation and emissions trading would work were left to Kyoto to be negotiated and resolved in subsequent meetings; in the years since the Protocol was completed, it became increasingly clear that this is an extremely complex issue, and an emissions trading system is not likely to be designed and implemented quickly. This has been a major element of the work on the Buenos Aires action plan, and remained far from resolved in the text introduced for negotiation in The Hague in November.

Another major "mechanism" for meeting obligations in the Protocol is provided by the establishment of a "clean development mechanism" (CDM), through which joint implementation between developed and developing countries would occur. The United States has pushed hard for joint implementation, and early proposals were formulated with the expectation the "JI" projects would be primarily bilateral. Instead, negotiations resulted in agreement to establish the clean development mechanism to which developed Annex I countries could contribute financially, and developing non-Annex I countries could then use certified emission reductions from such projects to contribute to their compliance with part of their emission limitation commitment. Emissions reductions achieved through this mechanism could begin in the year 2000 to count toward compliance in the first commitment period (2008-2012). Again, proposals on how this mechanism would operate are to be developed and negotiated under the Buenos Aires action plan. Like emissions trading, making the CDM operational appears likely to be a protracted and difficult process, given the increasing number of complexities emerging from the on-going work and discussions on how the CDM might work. However, at The Hague, how the CDM might operate became somewhat better articulated.

BUENOS AIRES ACTION PLAN

Although it had been expected just after the 1997 Kyoto conference that the November 1998 COP-4 meeting in Buenos Aires would resolve some of the more difficult issues left unresolved in Kyoto, it became clear during the year leading up to COP-4 that parties were far from agreement on all of these issues. Additional time for parties to analyze, negotiate, and work on these issues would be required. Therefore, the parties arrived in Buenos Aires with an agenda focused on formulating an "action plan" that would allow for the needed additional work to be done. It was decided that the work plan would be completed by the end of 2000, and would focus on the key issues, including the following:

- Rules and guidelines for the "market-based mechanisms" that allow flexibility to parties in meeting their obligations. These include emissions trading, joint implementation, and the Clean Development Mechanism (CDM). The list of critical issues to be considered include whether there should be quantified limits on how much of a country's emission reduction requirement could be met through these mechanisms, as argued by the European Union, or no quantified limit, as argued by the United States; "transparency" in making it possible to effectively track emission units; allocating risk in emissions trades – including the question of assigning liability, or responsibility, when emissions trading involves "false" credits; and key measurement, reporting and verification issues.
- Rules and procedures that would govern compliance, including provisions covering non-compliance and the treaty's commitments. This issue was left entirely open at Kyoto and remains one of the major challenges facing negotiators.
- Issues concerning development and transfer of cleaner, lower-emitting technologies, particularly to developing countries.
- Consideration of the adverse impacts of climate change and also the impacts of measures taken to respond to it, an issue of particular importance to developing countries, who argue the need for financial assistance in order to help them cope with these impacts.

Carbon sinks. Another issue under active negotiations and consideration by the parties outside the action plan itself is defining application of the concept of carbon sinks, including how to measure and verify the categories of carbon sinks. The scientific panel that provides analysis to the parties, the Intergovernmental Panel on Climate Change (IPCC), conducted a comprehensive study on land use, land-use change, and forestry activities to identify their roles as carbon sinks and deal with the measurement and verification issues related to them.

Following the release of this report, which indicated that a large amount of carbon could be stored in a variety of carbon sinks, including not only forests, but in soils, vegetation, grazing lands, etc., the United States made a comprehensive proposal for the COP-6 negotiations to broaden the scope of acceptable carbon sinks. The Kyoto Protocol accepts in principle that a nation's forests – management practices, reforestation or afforestation – may be included in the accounting of net greenhouse gas emissions and their reduction. This is important to the United States, as its large land area and extensive potential for greater absorption of carbon due to land management changes could greatly reduce the amount of emissions reductions needed from energy production. In a submission to the Secretariat of the UNFCCC, the United States proposed in late summer 2000 that elaboration at COP-6 of land use changes acceptable under the Protocol should also include soil carbon sequestration and vegetation.

Few decisions were reached, nor were they expected, on the more difficult issues outline in the Buenos Aires Plan of Action at the COP-5 meeting in Bonn, Germany, held October 25-November 24, 1999.

COP-6 NEGOTIATIONS

There were two inter-sessional negotiations sessions of the UNFCCC Subsidiary Bodies (the key mechanisms for considering Kyoto issues between COP meetings) in 2000, June 12-16 and September 11-15, dealing with the issues of the Buenos Aires work plan and attempting to fashion a negotiating text for final consideration at the November 13-24 COP-6 meeting. At the conclusion of the meeting in Lyon, it was reported that language on aspects of some issues had been agreed upon, but that the negotiating text had been expanded to some 200 pages, much of it "bracketed." [Brackets indicate that no agreement has been found on the language in brackets, and often several alternative possibilities are reflected in the brackets.] Thus, though there was negotiating text on most issues, disagreements remained on most key issues. Observers reported that political positions remained entrenched and little movement toward compromise appeared evident in Lyon. The end result was that many participants felt doubtful that many of the key issues could be completely resolved in November 2000, at COP-6.

As talks began at COP-6 in The Hague November 13, they centered initially on the "Buenos Aires Plan of Action" (BAPA), but evolved into a high-level negotiation over the major political issues. These included major controversy over the United States' proposal to allow credit for carbon "sinks" in forests and agricultural lands, satisfying a major proportion (between half and one-quarter, according to various versions of the U.S.

proposal) of the U.S. emissions reductions in this way; disagreements over consequences for non-compliance by countries that did not meet their emission reduction targets; and difficulties in resolving how developing countries could obtain financial assistance to deal with adverse effects of climate change and meet their obligations to plan for measuring and possibly reducing greenhouse gas emissions.

In the final hours of COP-6, despite some compromises agreed between the United States and some EU countries, the EU countries as a whole, reportedly led by Denmark and Germany, rejected the compromise positions, and the talks in The Hague collapsed. Jan Pronk, the President of the COP, suspended COP-6 without agreement, and it is expected that talks will resume at meetings in May/June 2001 in Germany, (a session now referred to as "COP-6 bis," or a continuation of the suspended COP-6 meetings in The Hague) and continue at the next COP-7, to be held in October 2001 in Marrakech, Morocco. Discussions between the EU and the "Umbrella group" that includes the United States, Canada. Japan, and Australia, were held in Ottawa, Canada, during the week of December 4 in order to salvage some of the agreement reached at the end of the talks in The Hague. However, the U.S. negotiators reported that these talks were "inconclusive" and the differences still in place. It is unclear whether additional talks will be held before the May meetings in Bonn, Germany, but informal discussion are likely to continue.

The political issues that remain most difficult and are most important to the United States, include:

Mechanisms, especially emissions trading: The main issue here is "supplementarity" – the position of the United States is that there should not be quantitative limits to the amount of emissions reductions that are allowed toward a country's obligations through emissions trading or joint implementation. The EU and others argue there should be such limitations, in order to force nations to take more extensive domestic action to reduce emissions. This issue is related to the commitment outline in the Kyoto Protocol that emissions trading should be "supplemental" to domestic action. The United States is supported by the "umbrella group" in which it is joined by New Zealand, Japan, Canada, Australia, Russia, Ukraine, Norway and Iceland. Another related issue is whether carbon sinks can be included in the Clean Development Mechanism (CDM) in which a contributing developed country can credit for actions to reduce emissions in developing countries. There are significant divisions on this issue not among developed countries, but among developing countries, as well.

Compliance issues: Decisions on how non-compliance with Protocol (and UNFCCC) obligations should be handled is a very controversial issue. The United States position is that there should be binding consequences, but these should be in the form of additional obligations in subsequent commitment periods, and not in the form of financial penalties. There was considerable disagreement at The Hague over whether financial penalties should be allowed. Agreement on binding consequences would probably require an amendment to the Protocol, which would be separately agreed to and separately ratified. Opponents to binding consequences include Japan, Russia, and Australia, who are concerned, among other things, about opening the Protocol to an amendment process. On the structure of a compliance regime, there was substantial agreement on a likely outcome: a single compliance body with two functions or branches – (1) to facilitate and assist compliance (mainly for developing countries), and (2) an enforcement function where decisions would be made on whether compliance violations have occurred and what consequences should be applied. Consequences under consideration, in addition to financial penalties, include losing access to mechanisms like emissions trading and/or subtracting from future allocations of allowable carbon emissions; however, the issue of what consequences may be agreed upon is apparently far from a final decision.

Land use and land use change and forestry (LULUCF): As noted above, the Kyoto Protocol accepts in principle that a nation's forests-management practices, reforestation or afforestation – may be included in the accounting of net greenhouse gas emissions and their reduction. The United States proposed at COP-6 that land use changes acceptable under the Protocol should also include soil carbon sequestration and vegetation. Major issues are how to attain precision in measuring absorption and release of carbon from land-based sources, how permanent such land use mechanisms would be, and the extent to which land use absorption counted by a nation is "additional" to business as usual.

The United States made a series of controversial proposals on how to count carbon sequestration, beginning with basically counting most of the carbon sequestration in its extensive forest cover toward its obligation. It subsequently revised this proposal and put forward a formula that included three parts: a first "interval" allowing up to 20 million tons of carbon to be counted at 100% for any country with forests absorbing that much; a second interval of a certain amount in which credit for a certain percentage would be allowed up to a certain threshold; then full credit for tons absorbed beyond the threshold (which would be historically determined in relation to baseline absorption amounts). The EU opposed the U.S. proposal, mainly on the issue of forests, and the extent to which a country like the United States would receive credits for a "business as usual" scenario that did no involve the harder emissions reductions from fuel sources and technological measures. When the United States put numbers to this proposal, the U.S. credits from carbon sinks appeared to represent about 125 million tons of carbon, against a likely need to reduce emissions by about 600 million tons of carbon to meet its commitment in 2008-2012. This was strongly opposed by the EU and other countries, and a stalemate over this issue, despite several revisions downward of the U.S. position and tentative acceptance of a much smaller amount by the EU, is thought to be a major factor in the collapse of the COP-6 negotiations at The Hague.

Developing country participation: The United States has been seeking additional commitments from developing countries in a series of informal discussions and consultations during the nearly three years since the Kyoto Protocol was completed in late 1997, but only Argentina and Kazakhstan have shown a willingness to make such commitments. Little willingness to do so has been shown by other developing countries. This issue is key to any consideration of ratification by the United States. The possibility that it would be discussed under an agenda item dealing with "adequacy of commitments" did not occur at The Hague.

A little over a week after the collapse of negotiations at The Hague, the EU and Umbrella Group countries met in Ottawa, Canada, to attempt to salvage some of the agreement reached in the waning hours of the COP-6 meeting. However, these talks were "inconclusive" according to the U.S. officials, and it is unclear when such informal talks might resume. COP-6, have been suspended at The Hague, will probably resume at the May/June meetings in Bonn, Germany, ("COP-6 bis") which were scheduled to be inter-sessional meetings leading up to COP-7, scheduled for October 2001 in Marrakech, Morocco.

ISSUES FOR CONGRESS

Ratification. For the United States to ratify the Protocol, the treaty must be submitted to the U.S. Senate for advice and consent, with a two-thirds majority vote in the Senate required for approval. If the United States does not ratify the treaty, it is not subject to its terms and obligations. President Clinton strongly supported the Kyoto Protocol, though criticizing it for not including commitments for developing countries. The United States signed the Protocol on November 12, 1998. The U.S. signature was criticized by several Members of Congress who oppose the treaty on a number of grounds, including questions about the scientific justification for it and about the likely economic impacts that might occur if the United States were to attempt to meet its emission reduction commitments in the treaty. In recognition of the opposition expressed in the Senate by S.Res. 98, which passed 95-0, to a Protocol that does not include requirements for emissions limitations by developing countries, President Clinton did not submit the treaty to the Senate for advice and consent, citing lack of meaningful developing country participation.

Oversight. Both the House and Senate have sent and continue to send delegations of Members to serve as observers on the U.S. delegation to Kyoto-related meetings. Supporters and opponents of the Protocol have been included in these delegations. A number of committees have held hearings on the implications of the Protocol for the United States, its economy, energy prices, impacts on climate change, and other related issues. While the Clinton Administration stated that it believed the treaty could be implemented without harm to the U.S. economy, and without imposing additional tax, a number of questions related to how its goals can be achieved and at what cost, continue to be of interest to Congress.

Legislation. If a treaty were to be sent to the Senate for consideration, legislation that might be required for its implementation would also typically be sent to the Congress by the Administration. Such legislation would not be likely unless the Kyoto Protocol were sent to Congress. President Clinton's proposal on climate change, announced in October 1998, included, among other things, a $5 billion package of tax credits and spending on research and development over 5 years to encourage energy efficiency and development of new lower emission technologies. In subsequent budget proposals, President Clinton offered an initiative over multiple years of $6.4 billion dollars for research and development and some possible tax incentives. A number of legislative proposals in the 105th and 106th Congresses – including bills, resolutions, and provisions in several appropriations bills enacted in the 106th Congress contained "riders that limit activities of the U.S. government that might be sent to advance the goals of the Kyoto Protocol prior to its consideration by the Senate. Others would provide for research priorities and some bills would provide for early credit for greenhouse gas emission reductions in the United States (see "Legislation" Section of the CRS Global Climate Change Electronic Briefing Book – [http://www.congres.gov/brbk/html/ebgcc1.html] and/or the legislation section of CRS Issue Brief 89005).

Chapter 6

GLOBAL CLIMATE CHANGE: MARKET-BASED STRATEGIES TO REDUCE GREENHOUSE GASES

Larry Parker

MOST RECENT DEVELOPMENTS

The November Kyoto Protocol negotiations at The Hague failed to come to agreement on the implementation for the three market mechanisms contained in the Protocol. The next negotiations are set for May 2001.

In late September 2000, Presidential candidate George W. Bush proposed a national energy plan that would include requiring utilities to reduce their carbon dioxide emission over a "reasonable" time frame in a manner similar to the current market-based acid rain reduction program. Few specifics, such as reduction targets or schedule, were included in the plan.

The environmental agenda laid out by Vice President Gore includes ratifying the Kyoto Protocol, and creating an Energy Security and Environmental Trust Fund to encourage among other things, utilities to clean up their older coal-fired electric generators. As noted, the Kyoto Protocol contains binding targets and timeframes for greenhouse gas reductions, and includes several market mechanisms to lower compliance costs.

BACKGROUND AND ANALYSIS

Certain gases emitted as a result of human activities may be affecting global climate. Most concern centers on the possibility that CO_2, along with other gases, could increase global temperatures, with subsequent effects on precipitation patters and ocean levels that could affect agriculture, energy use, and other human activities.

STATUS OF GLOBAL CLIMATE CHANGE ISSUE AND RESPONSE

The initial issue of whether the potential for global climate change poses a threat that justifies prompt action to curtail CO_2 and other so-called greenhouse gases remains actively debated-both domestically and internationally. Some view the risks as sufficiently grave and urgent to justify immediate action. Others are uncertain of the risks but believe that selected policies to reduce emissions can be justified for other reasons and would provide insurance if the risks were borne out; these other reasons include improved energy efficiency, reduce reliance on imported oil, and increase revenues. Still others caution that actions to reduce CO_2 and other greenhouse gases could disrupt the nation's economy and should not be undertaken unless further scientific evidence of risks becomes available.

Despite the uncertainties, however, scientists and policymakers have increasingly adopted the view that human activities are releasing greenhouse gases at rates that could affect global climate. As a result, initiatives are underway to address the issue, resulting in proposals for national and international programs to curtail emissions.

An agreement on a United Nations' Framework Convention on Climate Change was on the agenda at the U.N. Conference on Economic Development in Rio de Janeiro in June 1992. The United States was an early signatory to the agreement, which was approved by the Senate October 7, 1992. In April 1993, President Clinton directed the federal government to craft a plan that would stabilize U.S. greenhouse gas emissions at 1990 levels by the year 2000. However, in 1997 it is projected that the Unite States will not meet its voluntary commitment at Rio to stabilize greenhouse gas emissions at 1990 levels by the year 2000. Indeed, it is unclear when U.S. carbon emissions may stabilize. A November 1997 report by the Energy Information Administration estimates U.S. carbon emissions in the year 2020 will be 45% above their 1990 levels.

Meanwhile, the United States and other signatories to the Climate Change Convention prepared to meet in December 1997 in Kyoto, Japan, in an effort to conclude negotiations on a binding protocol for specific provisions to reduce emissions of greenhouse gas emissions. In October 1997, jus before a meeting in Bonn, Germany, a preliminary for the Kyoto meeting, the White House announced a new position on reducing greenhouse gases, calling for stabilization at 1990 levels by the years 2008-2012. This position was modified in December at Kyoto to be more flexible. The final protocol agreed to at Kyoto requires the United States to reduce emissions of six greenhouse gases (methane, nitrous oxide, hydro fluorocarbons, per fluorocarbons, sulfur hexafluoride) by 7% on average from 1990 levels over the period 2008*2012. In November 1998, the parties met in Buenos Aires to develop work plans for specific elements of the Kyoto Protocol, including the trading of emission reductions and the Clean Development Mechanism. The parties decided that these work plans should be completed by the year 2000. In November 1999 meeting in Bonn postponed decisions about emissions trading under the November 2000 meeting at The Hague. The meeting at The Hague failed to arrive at agreement on emissions trading, and further negotiations are scheduled for May 2001.

Thus, despite continuing uncertainties about the risks of global climate change, proposals for addressing it are going forward, and it is the content of those proposals rather than the issue of whether the problem is exigent that is the focus of this brief.

ESTIMATING COST IMPACTS OF CONTROLS

The potential cost of any CO_2 reduction program is of intense concern, not only because it would affect the costs of energy essential to the economy but also because the uncertainty of climate change suggests that the investments in controls could prove unnecessary. (While several gases can contribute to climate change and different proposals address various numbers of them, all the proposals would reduce emissions of CO_2, which is released in the combustion of fossil fuels and globally accounts for half or more of the potential global warming effect.)

Estimates of costs to reduce CO_2 emissions vary greatly, and focus attention on an estimator's basic beliefs about the problem and the future, rather than on simple, technical differences, in economic assumptions. These are summarized in Table1. None of these perspectives is inherently more "right" or "correct" than another; rather, they overlap and to varying degrees complement and conflict with each other. People hold to each of the lenses to some degree.

However, the differing perspectives lead to very different cost estimates. Figure 1 below shows a scatter-plot by World Resources Institute (WRI) of the predicted impacts from 162 estimates from 16 different economic models on the U.S. economy from a CO_2 abatement program. Although the size of the proposed CO_2 reduction and the time allowed to achieve it (not explicitly modeled in the WRI report) are critical factors in determining the costs and benefits of any reduction program, WRI found underlying modeling assumptions not related to policy decisions explained a significant amount of the difference in the estimates. For example, consistent with a "technological" view of the problem, models that assumed technological development of non-carbon substitutes for current fossil fuel use, along with increased energy and product substitution, had significantly less cost than models that assumed such advancements would not occur in a timely fashion. For example, a recent study by the American Council for Energy-efficient Economy (ACEEE) argues that carbon emissions could fall 10% below 1990 levels by 2010 with a net economic savings of $58 billion along with 800,000 new jobs. Such savings are assumed to come from new technology and market mechanisms to encourage cost-effective implementation strategies. Such a position presumes that technologies are available now, or will be very shortly, that can achieve these reductions cost-effectively.

Table 1. Influence of Climate Change Perspectives on Policy Parameters

Approach	Seriousness of Problem	Risk in developing mitigation program	Costs
Technology	Is agnostic on the merits of the problem. The focus is on developing new technology that can be justified from multiple criteria, including economic, environmental and social perspectives.	Believes any reduction program should be designed to maximize opportunities for new technology. Risk lies in not developing technology by the appropriate time. Focus on research, development, and demonstration; and on removing barriers to commercialization of new technology.	Viewed from the bottom-up. Tends to see significant energy inefficiencies in the current economic system that currently (or projected) available technologies can eliminate at little or no overall cost to the economy.
Economic	Understands issue in term of quantifiable cost-benefit analysis. Generally assumes the status quo is the baseline from which costs and benefits are measured. Unquantifiable uncertainty tends to be ignored.	Believes that economic costs should be examined against economic benefits in determining any specific reduction program. Risk lies in imposing costs in excess of benefits. Any chosen reduction goal should be implemented through economic measures such as tradable permits or emission taxes.	Viewed from the top-down. Tends to see a gradual improvement in energy efficiency in the economy, but significant costs (quantified in terms of GDP loss) resulting from global climate change control programs. Typical loss estimates range from 1-2% of GDP.
Ecological	Understands issues in terms of its potential threat to basic values, including ecological viability and the well being of future generations. Such values reflect ecological and ethical considerations; adherents see attempts to convert them into commodities to be bought and sold as trivializing issues.	Rather than economic costs and benefits or technological opportunity, effective protection of the planet's ecosystems should be the primary criterion in determining the specifics of any reduction program. Focus of program should be on altering values and broadening consumer choices.	Views cost from an ethical perspective in terms of the ecological values that global climate change threatens. Believes that values such as intergenerational equity should not be considered commodities to be bought and sold. Costs are defined broadly to include aesthetic and environmental values that economic analysis cannot readily quantify and monetize.

Figure 1. The Predicted Impacts of Carbon Abatement
on the U.S. Economy (162 estimates from 16 models)

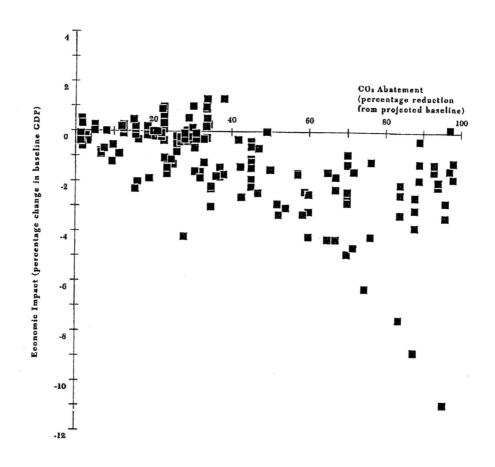

Likewise, consistent with an "ecological" perspective, models that include the benefits of air pollution damages and climate change damages averted by the CO_2 reduction estimated considerable less costs to the economy than models that did not include such benefits. The WRI report suggest that the cost profile of a CO_2 reduction program changes substantially if one includes the benefits of air pollution and climate change effect averted by controlling CO_2. The Administration's 1998 analysis of costs to comply with Kyoto estimates benefits from ancillary pollutants (SO_2, NO_x and fine particulates) at between $1.8 and $10.6 billion annually.

Consistent with an "economic" perspective, models that included policy approaches that encouraged efficient economic responses to CO_2 reductions, that include joint implementation schemes, and involved efficient recycling of any revenues from control strategies, significantly reduced costs over models runs that did not include such policy options... Like the technology perspective, economically efficient solutions assume that the program is implemented in such a way to permit the economy sufficient time to absorb the new price signals with minimal short-term constraints.

The uncertainty about the risk of climate change and the critical impact of assumptions about the nature of the problem effectively preclude predictions of the ultimate costs of reducing greenhouse gases. As a result, attention has focused on how to minimize costs by selecting the most economically efficient strategies to reduce CO_2 emissions. Traditionally, air pollution control programs have relied on various "command and control" regulatory approaches, including ambient quality and technology-based standards. But increasingly, economic efficiency concerns have been directed toward supplementing regulatory control with market-based mechanisms, including pollution taxes and tradable permits.

The tradable allowance system for SO_2 control in the acid rain program enacted in 1990 represents a significant step in this evolution of economic mechanisms. Acceptance of this system has led to calls for use of a similar system with other pollutants, including CO_2. A bill proposing a tradable permit-type system to being controlling CO_2 emissions was introduced during the 105th Congress. None has been introduced in this Congress.

MARKET-BASED MECHANISMS FOR REDUCING GREENHOUSE GASES

Proposals to use market mechanisms to implement greenhouse gas emission reductions have revolved around three approaches: tradable permits (as "allowances" and as "credits"), carbon taxes, and joint implementation. The protocol negotiated at Kyoto contains articles on emissions trading and joint implementation. These provisions are strongly supported by the Clinton Administration. In addition, some European countries have implemented or are considering carbon taxes to bring about greenhouse gas reductions in their countries.

Tradable Permits (Allowances)

A model for a tradable permit approach is the SO_2 allowance program contained in Title IV of the 1990 Clean Air Act Amendments. The Title IV program is based on two premises. First, a set amount of SO_2 emitted by human activities can be assimilated by the ecological system without undue harm. Thus the goal of the program is to put a ceiling or cap, on the total emissions of SO_2 rather than limit ambient concentrations. Second, a market in pollution rights between polluters is the most cost-effective means of achieving a given reduction. This market in pollution rights (or allowances, each of which is equal to one ton of SO_2) is designed so that owners of allowances can trade those allowances with other emitters who need them or retain (bank) them for future use or sale. Initially, most allowances were allocated by the federal government to utilities according to statutory formulas related to a given facility's historic fuel use and emissions; other allowances have been reserved by the government for periodic auctions to ensure the liquidity of the market.

Conceptually, a CO_2 tradable permit program could work similarly. Some number of CO_2 allowances could be allocated, and a market in the allowances would permit emitters to sue, sell, buy, or bank them. However, significant differences exist between acid rain and possible global warming that may affect the appropriateness of a Title IV-type response to CO_2 control. For example, the acid rain program may involve up to 3,000 new and existing electric generating facilities that contribute two-thirds of the country's SO_2 and one-third of its nitrogen oxide (NO_x) emissions (the two primary precursors of acid rain). This concentration of sources makes the logistics of allowance trading administratively manageable and enforceable. However, CO_2 emissions are not so concentrated. Although over 955 of the CO_2 generated comes from fossil fuel combustion, only about 33% comes from electricity generation. Transportation accounts for about 33%, direct residential and commercial use about 12%, and direct industrial use about 20%. Thus, small dispersed sources in transportation, residential/commercial, and the industrial sectors are far more important in controlling CO_2 emissions than they are in controlling SO_2 emissions. This creates significant administrative and enforcement problems for a tradable permit program it if attempts to be comprehensive.

These concerns multiply as the global nature of the climate change issue is considered, along with other potential greenhouse gases. Article 3 of the protocol negotiated at Kyoto emphasizes that any international emissions trading should be supplemental to a country's domestic efforts, not a substitute for them.

Current SO_2 allowance trading plans between individual utilities do not shed much light on how well the existing allowance market will work over the long term. Some individual trades between utilities and EPA-sponsored auctions have been conducted, but the current level of trading activity has not established the viability of the marketplace. For a market to thrive, transactions must become sufficiently commonplace for an open, public market price to be established with limited bilateral negotiations. Based on the results of the EPA auctions conducted by the Chicago Board of Trade, allowance prices are considerably below that anticipated when the legislation was enacted. However, the four-year experience of the SO_2-allowance market is as yet insufficient to give much guidance on how well a CO_2-allowance market might work.

Tradable Permits (Credits)

As noted above, a tradable allowance involves future emissions. An allowance is a limited authorization to emit a ton of pollutant; allowances are allocated to an emitting facility under an applicable emission limitation at the beginning of a year. The facility decides whether to use, trade, or bank those allowances, depending on its emissions strategy. Then, at the end of the year, the agency compares an emitting facility's actual emissions with its available allowances to determine compliance.

A different approach to creating a tradable permit program is to use credits instead of allowances. A credit is created when a facility actually emits a pollutant at less than its allowable limit as defined in by the program. An example of this type of program is EPA's "Emission Reduction Credit program" (ERC) under the Clean Air Act. Under the ERC program, EPA requires that any credit created under a state program implementing emissions trading be "surplus, enforceable (by the state), permanent, and quantifiable." Thus, a state must certify the creation of the credit, unlike an allowance program, where allocation is dictated by a statutory or regulatory formula. Any CO_2 reduction credit program could build on EPA's and states' experience with the current emission reduction credit program.

The primary advantage of a credit program over an allowance program is that it does not discriminate against new sources. Allowance programs tend to allocate their allowances based on some historic baseline year. Those sources included in the baseline get their allowances free. Those future sources not included in the baseline have to pay either the older, existing sources to obtain allowances or to buy allowances at auction. With a credit program, sulfur credits can be created by any source, as the baseline is dictated by the emissions cap and yearly production, not a historical year. The disadvantage of such a system is that facility planning is very difficult, as operators do not know precisely what their permissible limit will be from year to year.

Carbon/CO_2 Emissions Tax

An alternative market-based mechanism to the tradable permit system is carbon taxes – generally conceived as a levy on natural gas, petroleum, and coal according to their carbon content, in the approximate ratio of 0.6 to 0.8 to 1, respectively. In the view of most economists, the most efficient approach to controlling CO_2 emissions would be a carbon tax. With the complexity of multiple pollutants and million of emitters involved in controlling CO_2, the advantages of a tax are self-evident. Imposed on an input basis, administrative burdens such as stack monitoring to determine compliance would be reduced. Also, a carbon tax would have the broad effect across the economy that some feel is necessary to achieve long-term reductions in emissions.

However in other ways, a tax system merely changes the forum rather than the substance of the policy debate. Because paying an emission tax becomes an alternative to controlling emissions, the debate over the amount of reductions necessarily becomes a debate over the tax level imposed. Those wanting large reductions quickly would want a high tax imposed over a short period of time. Those more concerned with the potential economic burden of a carbon tax would want a low tax imposed at a later time with possible exceptions for various events. Emissions taxes would remain basically an implementation strategy; policy determinations such as tax levels would require political/regulatory decisions. In addition, a tax system would raise revenues. Indeed, one argument for – or against – such a system would be that it is a tax that would raise revenues. The disposition of these revenues would significantly affect the economic and distributional impacts of the tax.

Other tax schemes to address global climate change are also possible. For example, the European Community (EC) has discussed periodically a hybrid carbon tax/energy tax to begin addressing CO_2 emissions. Fifty percent of the tax would be imposed on energy production (including nuclear power) except renewables; 50% of the tax would be based on carbon emissions. Some European countries have modified their energy taxation to fit the model discussed by the EC.

Currently, four European countries have carbon-based taxes. Finland imposed the first CO_2 tax in 1990 and modified it in 1994. The Finnish tax has two components: (1) a basic tax component to meet fiscal needs and (2) a combined energy/CO_2 tax component. For coal, peat, and natural gas, there is no fiscal component. The Netherlands also introduced a CO_2 tax in 1990, modified in 1992 to fit the EC model. It does include tax relief from the energy component of the tax for energy-intensive industries. Sweden introduced a CO_2 tax in 1991 on all fossil fuels, unless it is used in electricity production. In 1993, the tax scheme was modified to reduce its burden on industry. Finally, Denmark introduced a CO_2 tax in 1992 that covers fuel oil, gas, coal, and electricity (gasoline is taxed separately). Taxes paid by industry are completely reimbursed to the sector. Norway introduced a CO_2 tax in 1991 on oil and natural gas and extended it to some coal and coke use in 1992. However, there are many exemptions and the tax rate if not differentiated according to the carbon content of the fuels.

JOINT IMPLEMENTATION

Joint Implementation (JI) is an attempt to expand the availability of cost-effective CO_2 reductions into the international sphere through a variety of different activities. Basically, a developed country (where opportunities for reducing emissions are expensive) needing CO_2 reductions to meet its obligations under any international treaty could obtain reduction credits by financing emission reductions in another country, usually a developing country (where more cost-effective reductions are available). As generally conceived, the developed country financing the reductions and the developing country hosting the reduction project would split the achieved reductions between them in some previously agreed-upon manner. Joint Implementation is a keystone of U.S. climate change policy; it was subject to considerable debate at the Conference of Parties (COP) meeting in Berlin. These discussions resulted in agreement to implement JI in a pilot phase. Projects must be compatible with and supportive of national environmental and development priorities; accepted, approved, or endorsed beforehand by the Parties' governments; and have anticipated environmental benefits and projected financing fully articulated beforehand. Credits generated cannot be used to meet the Rio Treaty year 2000 target; credit for post-2000 targets was left to the meeting in Kyoto.

The focus of the U.S. JI effort is the U.S. Initiative on Joint Implementation (USIJI). Managed by a Secretariat cooperatively staffed by 8 federal agencies, the USIJI is a pilot JI program initiated by the Administration as part of its "Climate Change Action Plan" in 1993. Currently, there are about 26 projects in 11 countries that have received USIHI approval. The USIJI encourages U.S. industry to use its resources and technology to reduce greenhouse gas emissions and promote sustainable development. (Its website is [http://www.ji.org].)

The advantage of JI for developed countries is that it widens the options available to obtain necessary credits under any reduction program. This translates into lower costs to those countries, compared with their own domestic reduction activities. For the developed country, particularly where it does not have the resources to control emissions or protect sequestration areas, reductions or protection would occur more quickly than would otherwise be possible.

However, the disadvantages are also significant. A developed country may have to rely on another sovereign government to ensure compliance with part of its international commitment. Governments change, and policies change. If a new government chose to remove or shut down a pollution control device, the developed country might have little recourse but to look elsewhere for its necessary reduction. Particularly with sequestration projects that involve marketable commodities, such as trees, enforcement could be quite difficult. A tree's value as cooking or heating firewood for natives could easily exceed its value as a carbon sequester. In the long run, the enthusiasm with which a developing country may enforce agreements with respect to JI projects is unclear.

Indeed, developing countries could have significant economic incentives to abrogate. JI projects, particularly if they are viewed as constraining necessary development, or locking up a natural resource that the country would like to exploit. This incentive is further encouraged if the JI project is perceived as a developed country's project. The term "economic imperialism" has already been applied to JI projects by some opponents.

After much negotiation, the protocol agreed to at Kyoto contains provisions on joint implementation that generally follow the guidelines set up at Berlin. Because developing countries have no emission requirements to meet (unlike developed countries), the protocol sets up a clean development mechanism to promote sustainable development in them while providing emission reduction opportunities for developed countries. Participation is voluntary; benefits must be real, measurable, and long-term; reductions must be in addition to any normal activity. Operated under supervision of the COP, reductions achieved between 2000 and 2008 may be used to offset commitments in the 2008-2012 time period.

ISSUES

Cost-Effectiveness: Price versus Quantity

Proposed CO_2 reduction schemes present large uncertainties in terms of the perceived reduction needs and the potential costs of achieving those reductions. In one sense, preference for a carbon tax or tradable permit system depends on how one views the uncertainty of costs involved and benefits to be received. For those confident that achieving a specific level of CO_2 reduction will yield very significant benefits – enough so that even the potentially very high end of the marginal cost curve does not bother them – then a tradable permit program may be most appropriate. CO_2 emissions would be reduced to a specific level, and in the case of a tradable permit program, the cost involved would be handled efficiently, but not controlled at a specific cost level. This efficiency occurs because control efforts are concentrated at the lowest cost emission sources through the trading of permits.

However, if one is more uncertain about the benefits of a specific level of reduction – particularly with the potential downside risk of substantial control cost to the economy – then a carbon tax may be most appropriate. In this approach, the level of the tax effectively caps the marginal control costs that affected activities would have to pay under the reduction scheme, but the precise level of CO_2 achieved is less certain. Emitters of CO_2 would spend money controlling CO_2 emissions up to the level of the tax. However, since the marginal cost of control among millions of emitters is not well known, the overall effect of a given tax level on CO_2 emission cannot be accurately forecasted. Hence, a major policy question is whether one is more concerned about the possible economic cost of the program and therefore willing to accept some uncertainty about the amount of reduction received (i.e., carbon taxes) or whether one is more concerned about achieving a specific emission reduction level with costs handled efficiently, but not capped (i.e., tradable permits).

A proposal was floated by the Administration for a tradable permit program with a ceiling on the price of permits. If permit prices rose above a certain price, the government would intervene to control costs by selling more permits at a specific price. In essence, this would give the permit program the character of a carbon tax by controlling costs through a price "safety valve," while allowing quantity to increase to any level necessary to prevent price increases. Not surprisingly, environmental groups interested in protecting the emission limitations of any global climate change program have attacked the idea as a "target-busting escape clause." Industry groups have suggested that such a tradable permit program amounts to a tax.

Comprehensiveness

As suggested earlier, carbon emissions are ubiquitous. Much of the emissions come from the direct combustion of fossil fuels from small, dispersed sources such as automobiles, homes, and commercial establishments. For example, the 12% of emissions from the residential/commercial sector comes from such things as space heating/cooling (9.3%, oil and natural gas), water heating (1.5%, mostly natural gas), and appliances (1.2%, mostly natural gas). If one adds to these dispersed sources the 33% of emissions that come from direct combustion from automobiles (13.9%), trucks (11.2%), airplanes (4.5%), ships (1.8%), pipelines (0.6%), and railroads (0.8%), the number of individual sources runs into the millions; very small sources contribute almost half the emissions.

Assuming a carbon tax is assessed on an input basis (i.e., on the carbon content of the fuel), then the number of sources is largely irrelevant – the sources would get the correct price signal from the increased cost of their fuel. This is one of the primary strengths of the carbon tax scheme – it can be very comprehensive and potentially induce the necessary changes in individual as well as corporate behavior that could substantially reduce dependence on carbon emitting energy sources. In this sense, a carbon tax is not just a band-aid to reduce CO_2 emissions, but a program to reduce carbon intensiveness in the economy and in individual lifestyles.

For a tradable permit program, the numbers of sources can represent a substantial administrative and enforcement problem. One approach to making the situation more manageable would be to limit the scope of the trading system to domestic implementation strategies. As noted above, international emission trading is termed "supplemental" under the consolidated negotiating text. Likewise, the scope could be limited further by focusing the trading program on the electric utility sector. Another approach could be to limit the size of the source included in the trading program. Others could "opt-in," but their participation would be voluntary. Thus, direct combustion of fossil fuels in the residential, commercial, and industrial sectors (e.g., natural gas, home heating oil) would be indirectly encouraged by the program and use of CO_2 emitting electricity (particularly coal-fired electricity) discouraged. The transportation sector would be little affected (unless it chose to be).

Economic Impact

Obviously, the economic impact of either a tradable permit program or a carbon tax depends on the level of reductions desired and the timing of those reductions. Most of the studies on the economic impact CO_2 control programs have focused primarily on carbon taxes. This is not surprising as carbon taxes are easier to model than a tradable permit program. However, the uncertainty involved in these analyses is quite large; further work is necessary to reduce the current range of estimates.

For example, estimates of the carbon tax necessary to stabilize U.S. CO_2 emissions at their 1990 level by the year 2000 range form under $30 a ton to over $100 a ton. Economic assumptions that result in this range of estimates include: (1) carbon emissions growth assumptions in the absence of legislation; (2) responsiveness of the economy to the carbon tax in terms of increased energy efficiency, and (3) type of model employed. This uncertainty is compounded when attempts are made to estimate GNP effects of carbon taxes. Very small differences in GNP estimation techniques can result in large differences in projected impacts (particularly over the long term). Preliminary evidence indicates that the adverse effects of certain existing taxes or fund investment incentives to encourage economic growth (particularly through greater capital formation). Thus, the impact of a carbon tax on the economy would depend to some degree on how the government disposed of generated revenues. However, considerably more work is needed to define the economic consequences of a specific proposal to recycle revenues before much confidence can be put into the results. Of course, if one has a technological or ecological orientation, the assumptions resulting from those orientations can draw the economic assumptions discussed here.

The extent that economic analysis of carbon tax programs provides insight for a tradable permit program depends partially on the scope of the program, the options included, and the monitoring and transaction costs. If the government chose to sell its allowances at auction, rather than give them away (as is typical), the government would have revenue like a carbon tax to recycle or readdress perceived distortions in the current tax code. In June 2000, CBO released a study on the distributional effects of carbon trading programs. It concludes that if the government gave away carbon allowances to U.F. firms (as is typical for trading programs), the effects are regressive on households. If the allowances are sold at auction, the distributional effects would depend on the ultimate disposition of the revenue received from the sale. However, the carbon tax analysis does suggest that the price of a permit (and any revenues from the sale thereof) would be difficult to estimate with any precision at the current.

The specific effects of both a carbon tax and tradable permit program would depend on the specific levy (carbon tax) or allocation scheme (tradable permit) chosen. Experience with both tax code revisions and the allocation scheme under the new acid rain title suggest that regional, state, and sector-specific concerns could receive special treatment in these decisions. In addition, for a carbon tax, the allocation of revenue received could also be influenced by such concerns.

Equity

The climate change issue and CO_2 control raise numerous equity issues. In one sense, the concern about climate change is a concern about intergenerational equity – i.e., the well being of the current generation versus generations to come. On a global level, the issue also involves the North-South debate. Some industrialized Northern countries suggest that the lesser-developed Southern countries refrain from certain activities (such as clearing rain forests) that Southern countries feel are important for their economic growth. Southern countries often suggest that the Northern countries change their current unsustainable growth practices and assist the South in sustainable development. Some supporters of tradable permits have suggested that internationalization of the permit program could allow the wealthy countries to fund CO_2-reducing activities (preserving forest, improving efficiency, etc.) as a means of achieving cost-effective reductions and assisting developing countries (i.e., joint implementation). However, as noted above, monitoring the long-term efficacy of JI projects raises administrative issues. Some carbon tax proponents have suggested that a portion of collected revenue could be set aside for assisting developing countries. Percentages to be set aside and more generally the political acceptability of such a proposal are unclear.

Other equity questions include the regional distribution of costs under a tradable permit or carbon tax scheme. For example, an important impact of either a carbon tax based on the carbon content of fossil fuels or a tradable permit program would be the pressure for fuel shifts away from coal and toward gas. Other regions, such as fast growing areas in need of more energy and owners of "all electric" homes, among others, would likely be disproportionately hit by a CO_2 control scheme. In addition, people may be affected differently according to income class. These issues have not been sufficiently analyzed at the current time to draw firm conclusions.

LEGISLATION IN THE 106TH CONGRESS

In the 106th Congress, seven bills have been introduced to control CO_2 emissions. S. 1369, introduced by Senator Jeffords, and H.R. 2569, introduced by Representative Pallone, H.R. 2645, introduced by Representative Kucinich, H.R. 2900, introduced by Representative Waxman, H.R. 2980, introduced by Representative Allen, and H.R. 4861, introduced by Representatives Lazio and Boehlert, reduce the cap emissions of carbon dioxide from electric generating facilities beginning in 2005. For S. 1369, H.R. 2569, H.R. 2900, and H.R. 2980, a cap of 1.914 billion tons would affect all electric generating facilities rated at 15 megawatts (Mw) or higher. For H.R. 4861, the 1.914 billion ton cap would affect all electric generating facilities rated at 25 Mw or higher. For H.R. 2645, a cap of a.71 billion would apply. In contrast, S. 1949 would impose unit-by-unit requirements on facilities that are estimated by the sponsor to result in CO_2 emissions at a level of 1-1.35 billion tons within 10 years of enactment.

Two additional bills, S. 547 introduced by Senator Chafee and H.R. 2520 introduced by Representative Lazio, would encourage voluntary greenhouse gas reductions by providing entities who participate early reduction credits for their actions. Applicable to actions taken before the year 2008, these credits would be useable in any future domestic reduction program. In general, a company's early actions would be compared with a company-wide baseline to determine the appropriate credits received. Credits are also available for reduction actions taken overseas, and verifiable actions recorded under Section 1605 of the Energy Policy Act of 1992.

OTHER PROPOSALS

Administration's Position

On October 22, 1997, the President formally released his proposal for reducing greenhouse gas emissions in preparation for the international conference in Kyoto, Japan, in December. The proposal called for a reduction in current greenhouse gas emission to their 1990 levels by 2008-2012. A key element of the Administration's proposal was the important role assigned emissions trading and JI. The President called for a domestic and international emissions trading system and international JI to implement emission reduction requirements to become effective 5 years after the 2008-2012 time period. The delay in the system was to allow a decade of experience with these mechanisms before they are fully implemented on such a broad scale. As stated by the President, "…the United States strongly supports the inclusion in a new climate change agreement of two innovative, flexible mechanisms for reducing emissions: International emissions trading [and] … Joint Implementation." Citing the 1997 Economic Report of the President, the Administration argued that international emissions trading for CO_2 could lower costs by 50% compared with a purely domestic program.

This support for trading programs has continued after the Kyoto Conference. During the June 1998, preparatory talks in Bonn, Germany, the United States submitted proposed rules on emissions trading for the Buenos Aires' meeting in November 1998. The proposal specifically put no limit on how much of a nation's emission target could be met by purchasing offsets from abroad. This proposal contrasted with that of the European Union that called for a "concrete ceiling" on the amount of emission reductions purchased from abroad. European opposition complicates the Administration's efforts to promote the greatest flexibility possible for countries to comply with the requirements of the Kyoto Agreement. At the Buenos Aires meeting in November, the parties were unable to call to agreement on emissions trading, instead calling on work plans for trading and other subjects to be completed by the year 2000. During May 1999, the debate within the European Union concluded with a proposal to cap the percentage of a country's greenhouse gas reduction requirement that could be reduced by international emissions trading at 50%. At the June 1999 talks in Bonn on implementing the Kyoto Protocol, the European Union was unable to convince the United States, along with Canada, Australia, New Zealand, and other countries, that emissions trading should be capped. The November 1999 meeting of parties at Bonn postponed any decision on emissions trading under the November 2000 meeting at The Hague. The Hague meeting was no more successful in reaching agreement on emissions trading, and the next meeting is scheduled for May 2001.

The inaction on formal emission trading parameters contrasts with a November 1999, announcement by the World Bank to launch a Prototype Carbon Fund (PFC) to earn carbon credits under the Clean Development Mechanism (CDM). The Bank hoped its initiative would attract $75-$100 million from governments and private industry to invest in Third World carbon reduction projects consistent with the Kyoto Protocol. In April

2000, the World Bank announced that it had exceeded its goals with 15 companies from 4 countries along with 6 national governments contributing nearly $135 million to its new PFC.

The Administration sees substantial economic advantages to international trading under Kyoto. In March 1998, the Administration testified that a carbon trading program among some of the developed countries (Annex 1 countries) could reduce U.S. compliance costs under the Kyoto Agreement by an estimated 60-75% compared with a compliance strategy that allowed no trading. Full participation by developing countries in a trading program is estimated by the Administration to reduce U.S. compliance costs by an additional 55%. Details on this analysis were released in late July. Assuming both international and domestic trading of emission reductions, the analysis puts the costs of compliance at $7-$12 billion annually. Under the analysis, about 75% of the necessary reduction would be met by buying reductions from abroad rather than by domestic control measures.

The administration has also suggested that other market reforms occurring in the U.S., particularly electricity restructuring, would assist the country in achieving cost-effective greenhouse gas reductions. In the case of electricity restructuring, this position is arguable, as there are several cross currents involved. For example, the Administration argues that cost studies indicate that the lion's share of cost savings from restructuring will come from increased use of existing coal-fired capacity. If true, then the need for new (cleaner) generating capacity could be delayed by restructuring. Increased use of existing capacity would also result in increased CO_2 emissions. Similarly, if prices for electricity decline, electricity use is likely to increase and incentives to conserve electricity likely to decrease. Declining prices would also reduce incentives for new technologies, including renewable energy.

Administration Domestic Initiatives

During his State of the Union Address, President Clinton endorsed the development of an early action credit program to encourage industry to make carbon dioxide (CO_2) reduction as early as possible. (For further information, CRS Report RC30155, *Global Climate Change Policy: Domestic Early Action Credits*.) The President also announced that the Administration would be proposing a series of tax incentives to encourage technologies that emit no CO_2.

The environmental agenda laid out by Vice President Gore includes ratifying the Kyoto Protocol, and creating an Energy Security and Environmental Trust Fund to encourage, among other things, utilities to clean up their older coal-fired electric generators. As noted, the Kyoto Protocol contains binding targets and timeframes for greenhouse gas reductions, and includes several market mechanisms to lower compliance costs.

Republican Proposals

In late September 2000, Presidential candidate George W. Bush proposed a national energy plan that would include requiring utilities to reduce their carbon dioxide emission over a "reasonable" time frame in a manner similar to the current market-based acid rain reduction program. Few specifics, such as reduction targets or schedule, were included in the plan.

Chapter 7

GLOBAL CLIMATE CHANGE: REDUCING GREENHOUSE GASES – HOW MUCH FROM WHAT BASELINE?

John Blodgett and Larry Parker

REDUCING GREENHOUSE GASES

On 15 October 1992 the United States ratified the United Nations Framework Convention of Climate Change (UNFCCC), which entered into force 21 March 1994. By this action, the nation committed to "national policies" to limit "its anthropogenic emissions of greenhouse gases." The goal was to return "...these anthropogenic emissions of carbon dioxide and other greenhouse gases" at the "end of the decade" "to their 1990 levels."[1] This goal was voluntary, to "demonstrate that developed countries are taking the lead in modifying longer-term trends in anthropogenic emissions consistent with the objective of the Convention."[2]

Subsequently, in the Kyoto Protocol to the UNFCCC, the United States agreed to a legally binding commitment to reduce emissions of greenhouse gases.[3] However, because the Protocol has not been sent to the Senate for its advice and consent, this commitment at present is only prospective; but if the Senate were to approve the Kyoto Protocol, the United States would be formally agreeing that over the 5-year period 2008-2012, it would

---

[1] UNFCCC, Article 4, Commitments, sections 2(a) and (b).
[2] Ibid., section 2(a).
[3] On the Agreement, see Susan R. Fletcher, *Global Climate Change Treaty: The Kyoto Protocol*, CRS Report RS30692; on the science and policy of global climate change, see Wayne A. Morrissey and John R. Justus, *Global Climate Change*, CRS Issue Brief IB89005. See also CRS's electronic briefing book *Global Climate Change* at [http://www.congress.gov/brbk/html/ebgcc1.html].

reduce its average annual aggregate carbon-equivalent emissions of 6 gases by 7% below specified baseline years.[4] The gases and the baseline years are as follows.

Greenhouse Gas	Baseline Year
Carbon dioxide (CO_2)	1990
Methane (CH_4)	1990
Nitrous Oxide (N_2O)	1990
Hydro fluorocarbons (HFCs)	1995
Per fluorocarbons (PFCs)	1995
Sulfur Hexafluoride (SF_6)	1995

As of early 2001, the deadline for achieving the voluntary reductions agreed to in the UNFCCC had passed; and only half the time remains from the 1990 baseline year to the reductions for 2008-2012 contemplated by the Kyoto Protocol. Where does the United States stand in terms of meeting its UNFCCC commitment? And where does it stand in terms of meeting the goal that it would be committed to achieve *if* the U.S, were to ratify the Kyoto Protocol? What, in short, is the status of U.S. greenhouse gas emissions?

As laid out below, the short answers are that aggregate emissions of greenhouse gases at the end of the decade have trended upward and have not held to 1990 levels. Moreover, the aggregate U.S. emissions trends are projected to continue to rise in the future, whereas the Kyoto Protocol calls for emissions to decrease 7% below the baseline for the period 2008-2012. While emissions could decrease in the future, every year of increasing emissions means that sharper cuts would have to be made over a shorter timeframe if the goals of the Kyoto Protocol were to be met.

This report sets out the baseline emissions that the U.S. has established; reviews emissions since the baselines; indicates the goals and implied reductions called for y the UNFCCC and the Kyoto Protocol; and evaluates projections for meeting Kyoto emissions goals. In what follows, *figures for emissions are point estimates and rounded to millions of metric tons of carbon equivalents (MMTCE)*. As will be discussed, even historical data may be subject to adjustments, though typically small ones not exceeding 1 or 2%. The data for CO_2, which accounts for over 80% of domestic greenhouse gas emissions, is the most robust, being largely based on comprehensive fuel use data. Subsumed estimates and uncertainties[5] in projected emissions have greater effect the further into the future one looks. But even allowing for these imprecisions, the *trend lines*

---

[4] Technically, the net carbon-equivalent emissions of the 6 greenhouse gases for the 5-year period 2008-2012 are not to exceed 5 times 93% of the baseline emissions. Kyoto Protocol, Article 3(1). This is equivalent to the average annual *emission load* during the 5-year period being 7% below the baseline.

[5] In a DOE study, the authors state: "The presentation here and throughout this report is not intended to imply high precision, but rather is designed to facilitate comparison among the scenarios and to allow the reader to better track the results. An uncertainty range for each value would be preferred to our single-point estimates, but the analysis required to prepare such ranges was not possible given the available resources..." DOE, Interlaboratory Working Group, *Scenarios for a Clean Energy Future* (Oak Ridge, TN; Oak Ridge National Laboratory and Berkeley, CA; Lawrence Berkeley National Laboratory, 2000) (ORNL/CON-476 and LBNL-44029). [http://www.ornl.gov/ORNL/Energy_Eff/CEF.html].

between baselines and the UNFCCC and Kyoto goals can give a clear sense of the challenge facing the United States.

U.S. GREENHOUSE GAS EMISSIONS AND BASELINES

Pursuant to the United Nations Framework Convention on Climate Change, the United States has published "national inventories of anthropogenic emissions by sources and removals by sinks of all greenhouse gases not controlled by the Montreal Protocol, using comparable methodologies...agreed upon by the Conference of the Parties."[6] The original U.S. response to this requirement to report on emissions and to explain programs was the *Climate Action Report*.[7] These early data have been superseded by the Environmental Protection Agency (EPA)'s annual data.[8] The data from EPA's *Inventory of U.S. Greenhouse Gas Emissions and Sinks: 1990-1998* are shown in table 1. EPA's data represent the inventory submitted by the U.S. to the UNFCCC. (EPA's inventory for 1990-1999, issued in draft[9] in January 2001, revises some of the emissions figures, although the changes for total emissions are less than 1% for 1998. This CRS report uses the data from the published 1990-1998 report but occasionally refers to data for 1999 from the draft, which should be considered preliminary.) The Energy Information Administration (EIA) also publishes greenhouse gas emissions data.[10] Because of methodological differences, EPA and EIA data vary slightly – EPA's CO_2 emissions are slightly lower than EIA's, but its figure for total greenhouse gas emissions for 1998 is about 1% higher than EIA's.

The EPA data provide the baselines for UNFCCC and the Kyoto Agreement, as shown in table 2. For the UNFCCC commitment, the baseline would be the 1990 emissions, or 1,649 MMTCE; for the Kyoto Agreement, the baseline would be 1,655

---

[6] UNFCCC, Article 4, section 1(a) and Article 12, section 1(a).

[7] Department of State, *Climate Action Report: 1997 Submission of the United States of America Under the United National Framework Convention on Climate Change*, (Washington, DC: U.S. Govt. Print. Off.), p. 111.

[8] EPA, "U.S. Emissions Inventory – 2000," *Inventory of U.S. Greenhouse Gas Emissions and Sinks: 1990-1998)*, April 2000, EPA 236-R-00-001, p. ES-4. [http://www.epa.gov/globalwarming/publications/emissions/us2000/index.html].

[9] EPA, *Inventory of U.S. Greenhouse Gas Emissions and Sinks: 1990-1999*. [http://www.epa.gov/globalwarming/publications/emissions/us2001/index.html].

MMTCE, since the baseline years for HFCs, PFCs, and SF_6 can be 1995. By definition, sinks are *not* included in calculating the baseline.

[10] EIA, *Emissions of Greenhouse Gases in the United States 1999*, (October 2000), DOE/EIA-0573(99). [www.eia.doe.gov/oiaf/1605/ggrpt/index.html].

Table 1: U.S. Greenhouse Gas Emissions, 1990-1998

Gas	1990	1991	1992	1993	1994	1995	1996	1997	1998
CO_2	1,340	1,326	1,350	1,383	1,405	1,416	1,466	1,486	1,494
CO_2 (sinks)[a]	(316)	(316)	(316)	(213)	(212)	(212)	(211)	(211)	(211)
CH_4	178	178	179	179	182	184	183	184	181
N_2O	108	110	113	114	122	119	122	122	119
HFCs, PFCs, SF_6	23	22	24	24	25	29	34	35	40
Total emissions	1,650	1,636	1,667	1,700	1,733	1,748	1,804	1,828	1,835
Net emissions	1,333	1,320	1,350	1,487	1,521	1,537	1,593	1,617	1,624

[a]Land-use changes and forestry sinks that sequester carbon; included in net emissions total only.

Source: EPA, "U.S. Emissions Inventory – 2000," *Inventory of U.S. Greenhouse Emissions and Sinks: 1990-1998* April 2000.

The emissions baselines shown in table 2 are not immutable. Each annual report includes updating based on methodological and data revisions, although such changes are usually small. (If EPA's draft figures for 1990-1999 hold, the baselines would each increase by about 10 MMTCE.[11]) The criteria for calculating emissions agreed upon by the Conference of Parties hinge on both current technical knowledge and on policy judgments.[12] New technical information could change factors, for example concerning calculation of greenhouse gas equivalents; and policy judgments could be adjusted, for example concerning the timeframe for calculating effects. In addition, a few technical issues remain unresolved, for example in assigning emissions from fuels burned in international travel – although any impact of such changes would probably be modest, perhaps 1 to 2%.

Table 2. U.S. Baseline Year Greenhouse Gas Emissions

Greenhouse Gas	Baseline Year		Emissions (MMTCE)	
Carbon dioxide (CO$_2$)	1990		1,340	
Methane (CH$_4$)	1990		178	
Nitrous Oxide (N$_2$O)	1990		108	
Hydro fluorocarbons (HFCs)	1990 UNFCC	1995 Kyoto	23 UNFCC	29 Kyoto
Per fluorocarbons (PFCs)				
Sulfur Hexafluoride (SF6)				
Total			1,649	1,655

Source: EPA, "U.S. Emissions Inventory – 2000," *Inventory of U.S. Greenhouse Gas Emissions and Sinks: 1990-1998),* April 2000, EPA 236-R-00-001, p. ES-4.

THE U.S. EMISSIONS GOAL

Under the UNFCCC, the U.S. committed to the voluntary goal of holding greenhouse gas emissions at the end of the 1990s decade to the 1990 levels. If the U.S. had met this goal, its greenhouse gas emissions for 2000 would have been, 1,649 MMTCE. However, in 1999, based on EPA's draft report of greenhouse gas emissions data, US. Emissions were 1,840 MMTCE (not counting sinks). These figures indicate that in 1999, the U.S. was exceeding its UNFCCC greenhouse gas emissions commitment by approximately 190 MMTCE, or about 10% (again, ignoring sinks).

The reductions that would be required of the U.S. to meet its commitment under the Kyoto Protocol – *if* the U.S. were to accede to the Agreement – will be the difference between 5 times 93% of the 1,6555 MMTCE baseline and what would be "business as usual emissions" for the period 2008-2012. This goal is 1,539 MMTCE on average per year. To calculate reductions necessary to meet future targets, emissions must be projected.

---

[11] EPA, *Inventory of U.S. Greenhouse Gas Emissions and Sinks: 1990-1999* [draft].

PROJECTING EMISSIONS TO 2008-2012

In the case of the Kyoto Agreement, reductions needed would be the difference between projected "business as usual" emissions in 2008-2012 and the annual average goal of 1,539 MMTCE. Projecting greenhouse gas emissions depends on computer modeling of economic growth and activity, with special attention to variables affecting fossil fuel combustion. The modeling also depends on assumptions about energy policy directions and the outcome of unresolved issues in how the Kyoto goal is to be met.

For example, the major source of CO_2 emissions, fossil fuel combustion, is influenced by overall economic activity and growth as well as by such energy policy considerations as development of non-carbon based substitutes, the rate of adoption of energy efficient technologies, and the rate of retirement of nuclear facilities, among others. These factors are difficult to predict in the absence of a discrete climate change policy. Also affecting the calculation of reductions is an unresolved Kyoto issue: while the Protocol allows sinks to be counted toward required reductions, how much they can count was left to future negotiations. This was a sticking point in the Sixth Conference of Parties held in The Hague in November 2000. The resolution of this issue has the potential for changing domestic fuel-related reductions required by as much as 10% or more.

The Climate Action Report Projections

The 1997 *Climate Action Report (CAR)* projects greenhouse gas emissions for the years 2000, 2005, 2010, and 2020. For evaluating the impact of the Kyoto Protocol, the 20120 projection falls in the middle of the target period. The projections presumed continued funding support for the Administration's Climate Change Action Programs comparable to the 1997 levels approved by Congress.[13] The report discussed uncertainties,[14] but did not indicate potential error bars for these point estimates.

Base on the *CAR* projection that emissions will be 1,946 MMTCE in 2010, the average reduction necessary for the U.S. to meet its Kyoto commitment would be 462 MMTCE per year, or 23.7% below the "business as usual" emissions projected. This calculation does not include any net, human-induced change in U.S. carbon sinks from afforestation, reforestation, and deforestation that could affect the 2010 reduction estimate.

---

[12] The Kyoto Protocol requires that further studies of greenhouse gas emissions and removal be undertaken and incorporated in any future commitments for reducing greenhouse gases beyond the 2008-2012 target.

[13] The *Climate Action Report* discusses variable affecting projections, pp. 117-124.

[14] The *Climate Action Report* discusses uncertainties, pp. 127-130.

The EIA Projections for CO_2

EIA's *Annual Energy Outlook* series includes CO_2 emissions projections.[15] Although these projections do not include some adjustments made in the national inventory of greenhouse gas emissions for the UNFCCC (e.g., bunker fuels used in international transport), such adjustments would constitute only a minor part of the overall projection – perhaps 1-2%. Given that CO_2 comprises about 82% of greenhouse emissions, these projections account for a large part of expected emissions.

Figure 1 depicts CO_2 emissions and projections. The solid squares and circles represent historical emissions as reported by EPA and EIA, respectively; the solid triangles show *CAR* estimates (1990, 1995) and projections (2000-2010); and the open circles show EIA's reference case forecast (2000-2010), with CRS estimated adjustments.[16]

Projected Total Emissions

To project aggregate emissions, it is possible to combine the following, in what might be termed a "Composite" Projection:

- The (adjusted) EIA forecast for CO_2 (figure 1),
- The *CAR* forecast for CH_4 which seems to be tracking fairly closely to actual data in the 1990s – EPA's preliminary 1999 figure of 167 MMTCE represents a drop that parallels EIA's and *CAR's* (see figure 2),
- The *CAR* forecast for HFCs, PFCs, and SF_6, emissions of which are trending upward as expected (see figure 3), and
- A re-estimated N_2O projection (actual 1990s data show that the relatively flat *CAR* projections is much too low), based on a flat projection that approximately averages EPA and EIA data, i.e., 110 MMTCE per year (see figure 4).

---

[15] EIA, *Annual Energy Outlook 2001* (Dec. 2000), DOE/EIA-0383(2001).

[16] Based on historical data from EIA, *Emissions of Greenhouse Gases in the United States 1999* (October 2000) DOE/EIA-0573(99), Table 4; the adjustments approximate those necessary to conform EIA data to UNFCCC methodology for computing emissions. Adjustments for fuels used in territories military bunker fuels, and international bunker fuels are estimated at – 17 MMTCE per year, and non-fuel sources are estimated to increase at a rate of 2% per year over 31.7 MMTCE in 1999. For 2010, these adjustments add just over 1% to the total projected in EIA's reference case (1,831 versus 1,809 MMTCE).

Alternatively, one could extrapolate from EPA's total emissions by assuming that the 1.2% annual growth rate from 1990 to 1998[17] will continue as "business as usual" through 2010. The result of this extrapolation comes out slightly less (about 3%) than the sum of the composite projection above.

---

[17] EPA, "U.S. Emissions Inventory – 2000," *Inventory of U.S. Greenhouse Gas Emissions and Sinks: 1990-1998* (April 2000) EPA 236-R-00-001, p. ES-2.

Figure 1. U.S. Emissions of CO₂: Historical (1990s) and Projected (to 2010) (MMTCE)

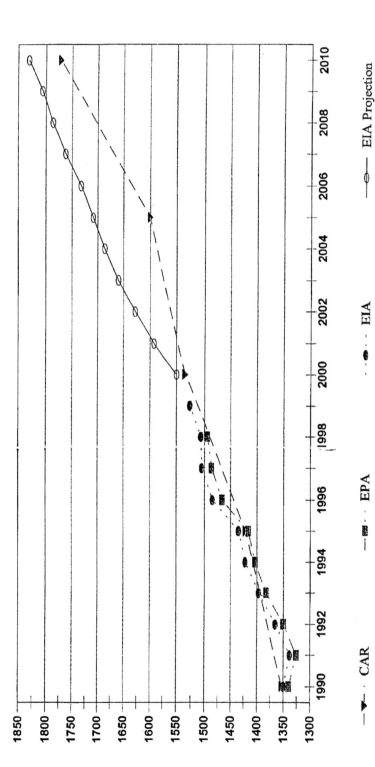

Sources:

1. Department of State, *Climate Action Report: 1997 Submission of the United States of America Under the United National Framework Convention on Climate Change* (Washington, DC: U.S. Govt. Print. Off.).

2. EPA, "U.S. Emissions Inventory – 2000," *Inventory of U.S. Greenhouse Gas Emissions and Sinks: 1990-1998* (April 2000) EPA 236-R-00-001.

3. EIA, *Emissions of Greenhouse Gases in the United States 1999* (October 2000) DOE/EIA-0573(99) Table ES2.

4. EIA, *Annual Energy Outlook 2001* (Dec. 2000), DOE/EIA-0383(2001), Appendix Tables, Table 19.

Figure 2. U.S. Emissions of CH₄ (MMTCE)

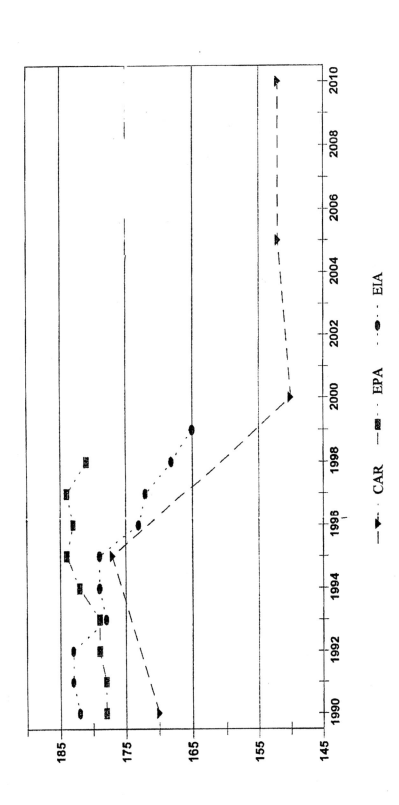

Sources:

1. Department of State, *Climate Action Report: 1997 Submission of the United States of America Under the United National Framework Convention on Climate Change* (Washington, DC: U.S. Govt. Print. Off.).

2. EPA, "*U.S. Emissions Inventory – 2000,*" *Inventory of U.S. Greenhouse Gas Emissions and Sinks: 1990-1998* (April 2000) EPA-236-R-00-001.

3. EIA, *Emissions of Greenhouse Gases in the United States 1999* (October 2000) DOE/EIA-0573(99) Table ES2.

Figure 3. U.S. Emissions of HFCs, PFCs, SF$_6$ (MMTCE)

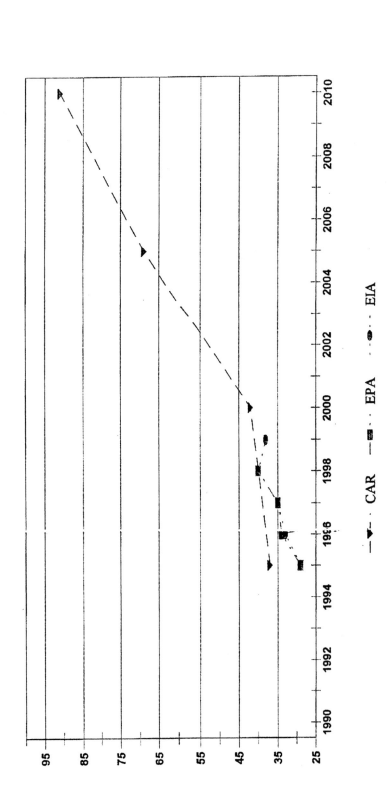

Sources:

1. Department of State, *Climate Action Report: 1997 Submission of the United States of America Under the United National Framework Convention on Climate Change* (Washington, DC: U.S. Govt. Print. Off.).

2. EPA, *"U.S. Emissions Inventory – 2000," Inventory of U.S. Greenhouse Gas Emissions and Sinks: 1990-1998* (April 2000) EPA-236-R-00-001.

3. EIA, *Emissions of Greenhouse Gases in the United States 1999* (October 2000) DOE/EIA-0573(99) Table ES2.

Figure 4. U.S. Emissions of N₂O (MMTCE)

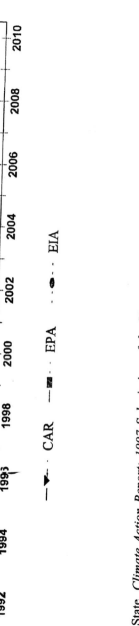

─▼─ CAR   ─■─ EPA   ⋯●⋯ EIA

Sources:

1. Department of State, *Climate Action Report: 1997 Submission of the United States of America Under the United National Framework Convention on Climate Change* (Washington, DC: U.S. Govt. Print. Off.).
2. EPA, "*U.S. Emissions Inventory – 2000," Inventory of U.S. Greenhouse Gas Emissions and Sinks: 1990-1998* (April 2000) EPA-236-R-00-001.
3. EIA, *Emissions of Greenhouse Gases in the United States 1999* (October 2000) DOE/EIA-0573(99) Table ES2.

Figure 5. U.S. Aggregate Emissions of Six Greenhouse Gases: Historical (1990) and Projected (to 2010) (MMTCE)

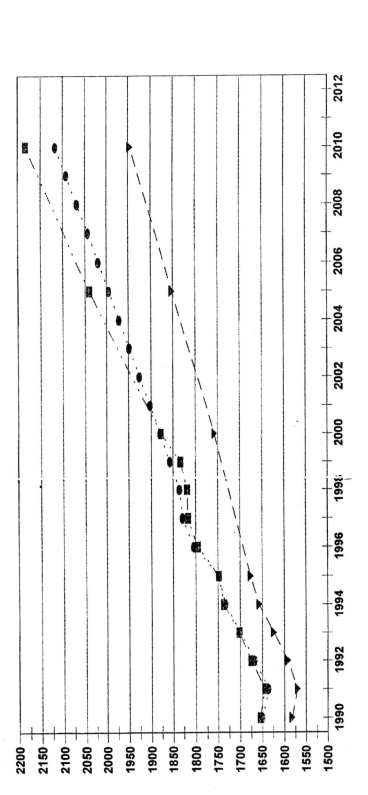

Sources:

1. Department of State, Climate Action Report: 1997 Submission of the United States of America Under the United National Framework Convention on Climate Change (Washington, DC: U.S. Govt. Print. Off.).

2. EPA, "U.S. Emissions Inventory – 2000," Inventory of U.S. Greenhouse Gas Emissions and Sinks: 1990-1998 (April 2000) EPA-236-R-00-001.

3. EIA, Emissions of Greenhouse Gases in the United States 1999 (October 2000) DOE/EIA-0573(99) Table ES2.

4. EIA, Annual Energy Outlook 2001 (Dec. 2000), DOE/EIA-0383(2001), Appendix Tables, Table 19.

Figure 5 graphically depicts the result of these projections, together with the goals of the UNFCCC for 2000 and of the Kyoto Protocol for 2008-2012. Based on the composite projection, which uses the most recent EIA projections for CO_2, the average annual gross reduction in greenhouse gases required to meet the Kyoto Protocol commitment would be 645 MMTCE for 2008-2012, a reduction of about 30% (ignoring sinks). Base on the extrapolation from EPA's emissions data, the gross reduction required to meet this obligation would be 578 MMTCE, a reduction of about 27% (also ignoring sinks). These figures bracket the conventional guesstimate that the Kyoto Protocol would require an annual 600 MMTCE reduction during 2008-2010.

These projected reduction requirements are gross estimates and do not take sinks into account. As previously noted, the baseline may be revised, at least slightly. More importantly, such projections depend on assumptions about economic trends as well as about policy actions at the local, domestic, and international levels. However, whatever the assumptions, the trend in total emissions over the past decade is clearly upward, while the Kyoto Protocol goal implies declining emissions levels over the next decade.

Variable Affecting Any Reduction Requirement

If the U.S. were to commit to reducing its emissions of greenhouse gases – whether because it acceded to the Kyoto Protocol, or for any other reason – several economic and policy variables would come into play that would affect the magnitude, costs, location, and parties responsible. Those that would have the greatest potential impact on emissions and which are topics of active policy deliberations include:

- Economic and energy policies, including the development and adoption of more energy efficient technologies, which can be affected by government programs;
- The accounting for CO2 sinks – a factor acknowledged by the UNFCCC and the Kyoto Protocol, but the extent to which they would be counted as contributing and reductions has not yet been agreed upon; and
- The use of any emissions trading mechanisms, which would be allowed under the Kyoto Protocol but have not yet been formally agreed upon. Trading would not change any reduction requirement, but it would affect who would actually undertake reductions and the costs.

Economic Variable/Energy Efficient Technology Market Penetration. Besides its reference case (baseline, business as usual) scenario, the *Annual Energy Outlook* 2001 projects CO_2 emissions from energy use under a series of alternative scenarios. High and low comparison cases are presented for economic growth and for oil prices, plus a series of "side cases" that include market penetration of energy efficient technologies (integrated technology), nuclear generation, electricity demand, electricity sector fossil technology, high renewable energy, oil and gal technological progress, and oil and gas resources.

The three scenarios that most affect CO_2 emissions are economic growth, integrated technology, and electricity demand, as show in table 3. A low economic growth assumption reduces projected emissions by 59 MMTCE, while the high economic growth assumption increases projected emissions by 74 MMTCE. Assuming greater penetration of new, energy efficiency technologies reduces projected emissions, while higher electricity demand increases projected emissions. However, these divergences from the reference case are relatively small; the largest, a 75 MMTCE reduction in emissions from high technology, represents about 3.55 of the composite projection shown in figure 5. However, this would represent about 12% of the reduction required to meet the Kyoto Agreement goal.

The role of the federal government in promoting the development and use of new, energy efficient technologies has been a key, continuing policy issue. Not only does the availability of such technology have implications for the magnitude of emissions, but also for the costs of reducing them. EPA and the Department of Energy (DOE) have underway a number of programs to foster the development and deployment of energy efficient technologies.[18]

Table 3. Impact of Economic Assumptions on Projections of CO_2 Emissions

Case Comparisons	Total CO_2 Emissions from Fuel Use, 2010 (MMTCE)		
	Low economic growth	Reference case	High economic growth
Economic Growth	1,750	1,809	1,883
	2001 Technology	Reference case	High Technology
Integrated Technology	1,866	1,809	1,734
		Reference case	High Demand
Electricity Demand		1,809	1,865

Source: EIA, Emissions of Greenhouse Gases in the United States 1999 (October 2000) DOE/EIA-0573(99). [www.eia.doe.gov/oiaf/1605/ggrpt/index.html]

Some studies suggest that new energy efficient technologies are available and could be deployed more quickly than generally assumed if appropriate policies were instituted. For example, the most recent DOE study, commonly called the "New 5-Lab Study," shows that energy efficiency gains in the transportation, industry, commercial, and residential sectors could reduce emissions from the "business as usual" scenario.[19]

The "business as usual" scenario in this study is very similar to EIA's reference case, though it projects slightly smaller emissions in 2010 (1,769 MMTCE from fossil fuel

[18] EPA, "U.S. Emissions Inventory – 2000," *Inventory of U.S. Greenhouse Gas Emissions and Sinks: 1990-1998* (April 2000) EPA 236-R-00-001, p. ES-2.

[19] DOE, Interlaboratory Working Group, *Scenarios for a Clean Energy Future* (Oak Ridge, TN; Oak Ridge National Laboratory and Berkeley, CA; Lawrence Berkeley National Laboratory, 2000) (ORNL/CON-476 and LBNL-44029). [http://www.ornl.gov/ORNL/Energy_Eff/CEF.html]

combustion, compared to EIA's most recent projection of 1,809). The study compares "moderate" and "advanced" scenarios "that are defined by policies that are consistent with increasing levels of public commitment and political resolve to solving the nation's energy-related challenges." Policies examined include "fiscal incentives, voluntary programs, regulations, and research and development."[20]

Under the "moderate scenario," energy efficiency is improved through such policies as expanded labeling, new efficiency standards, tax credits, and cost-shared R&D; renewable energy grows more rapidly than in the "business as usual" scenario, and a higher proportion of nuclear power is retained. Under the "advanced scenario," which has more aggressive demand-and supply-side policies and a doubling of R&D, a federal-sponsored carbon trading system is announced in 2002 and implemented in 2005 with a clearing equilibrium price of $50 per ton of carbon.[21] The results of this analysis are shown in table 4.

Table 4. Impact of Technology/Efficiency Assumptions on Projections of CO_2 Emissions

Case Comparison	Total CO_2 Emissions from Fuel Use, 2010 (MMTCE)
"Business as Usual"	1,769
Moderate Scenario	1,684
Advanced Scenario	1,467

Source: DOE, Interlaboratory Working Group, *Scenarios for a Clean Energy Future* (Oak Ridge, TN; Oak Ridge National Laboratory and Berkeley, CA; Lawrence Berkeley National Laboratory, 2000) (ORNL/CON-476 and LBNL-44029), Table 1.8, p. 1.18.

If these emissions reductions were achieved, they would represent savings of 4% (moderate scenario) to 14% (advanced scenario) over the composite projection (figure 5), based on EIA's reference case for CO_2 emissions. However, the advanced scenario would achieve about 30% of the Kyoto reduction goal. These reductions would not be sufficient to return U.S. CO_2 emissions to their 1990 levels, much less meet the Kyoto target.

Carbon Sequestration Variables. A country's contribution to greenhouse gases consists not only of direct emissions but also by carbon sinks – processes that remove and sequester carbon from the atmosphere. Activities that effect sinks include farming and forestry practices. For example, a positive net growth of trees removes carbon from the atmosphere; clearing forests typically releases carbon. Table 1, *U.S. Greenhouse Gas Emissions, 1990-1998,* includes figures for carbon sequestration from land-use activities and forestry. For 1998 EPA calculates the U.S. carbon "sink" at 211 MMTCE, the difference between "Total emissions" and "Net emissions."

The UNFCCC states that signatory nations shall commit to "promote sustainable management, and promote and cooperate in the conservation and enhancement, as

---

[20] Ibid., p. 1.4.
[21] Ibid., pp. 1.6-1.7.

appropriate, of sinks and reservoirs of all greenhouse gases…,including biomass, forests and oceans as well as other terrestrial, coastal and marine ecosystems" (Article 4(1)(d)).

The Kyoto Protocol also would provide that sinks can be taken into account in calculating a nation's emissions and its reduction obligation. "The net changes in greenhouse gas emissions from sources and removals by sinks resulting from direct human-induced land-use change and forestry activities, limited to afforestation, reforestation, and deforestation since 1990, measured as verifiable changes in stocks …shall be used to meet" the 2008-2012 commitments (Article 3(3)). In general, then, a net increase in human-induced carbon sequestration from forestry practices between 1990 and 2008-2012 would be subtracted from emissions during the period, thereby reducing the amount of actual emissions that will have to be curtailed. Conversely, net negative sequestration from forestry practices would be added to the emissions that will have to be reduced. Just how this calculation would be done is not prescribed in the Protocol, which states that revised methods of accounting for "removals in the agricultural soil and land-use change and forestry categories" would be a topic of the November 1998 Conference of Parties and could be applied in meeting the 2008-2012 commitment, if the activities took place after 1990 (Article 3(4)). However, the 1998 Conference at The Hague, where disagreements on how much carbon sequestration could be counted toward a nation's reduction obligations ultimately proved irreconcilable. Observers at the Conference report that during discussions of the maximum amount of sequestration that could be counted against a nation's reduction requirement, figures ranged from 50 MMTCE to 125 MMTCE. An allowance of 50 MMTCE would be about 8% of the approximately 600 MMTCE that the U.S. would have to reduce on an annual average if it were to accede to the Kyoto Protocol, while 125 MMTCE would account for over 20%.

Emissions Trading. The Kyoto Protocol provides for emissions trading mechanisms[22] that can be used to "supplement" domestic reductions. That is, the U.S. could achieve its Kyoto goal not only by reducing its domestic emissions, but also by reducing emissions from any certified source in a manner that results in the reduction requirement being met. Trading does not actually reduce a nation's reduction requirement, but it does allow it contract for and to count reductions elsewhere that are presumably cheaper to achieve than domestic ones.

The Protocol establishes three mechanisms that expand the inventory of reduction opportunities:

Under Article 4, the Protocol, authorizes Annex I Parties[23] "to fulfill their commitments under Article 3 jointly" – in effect, this allows one country to emit greenhouse gases in excess of its commitment to the degree another country's emissions are lower than its commitment similar to an emissions bubble. Parties proposing joint fulfillment of commitments must formally notify the secretariat of their intent when they ratify or otherwise approve the Protocol.

---

[22] *Global Climate Change: Market-Based Strategies to Reduce Greenhouse Gases*, CRS Issue Brief IB97057.

[23] Annex I Parties – listed in an Annex to the 1992 U.N. Framework Convention on Climate Change (FCCC) – include the "developed" nations and the former soviet economies; the U.S. in an Annex I Party. Each Annex I Party that has an assigned target for greenhouse gas emissions is listed, with its target, in Annex B of the Kyoto Protocol.

Under Article 6 of the Protocol, any Annex I Party "may transfer to, or acquire from any other such Party emission reduction units resulting from projects aimed at reducing anthropogenic emissions by sources or enhancing anthropogenic removals by sinks of greenhouse gases in any sector of the economy..." The Protocol spells out conditions for such emissions trading, including a requirement that if only "be supplemental to domestic actions for the purposes of meeting commitments..." The November 1998 Conference of Parties in Buenos Aires was to elaborate on guidelines for implementing this article with respect to verification and reporting. This issue was postponed until the 2000 Conference at The Hague, which was unable to reach agreement on guidelines.

Under Article 12, the Protocol defines a "clean development mechanism" by which Annex I countries can gain credit for post-2000 emissions reductions achieved by assisting non-Annex I countries in sustainable development activities that reduce emissions or enhance carbon sinks. Such "joint implementation" emissions reductions must be real and measurable; the November 1998 Conference of the Parties was to "elaborate modalities and procedures with the objective of ensuring transparency, efficiency and accountability through independent auditing and verification of project activities." The November 1998 COP postponed decisions on this and the other trading mechanisms until the 2000 conference at The Hague, but, again, the Parties were unable to come to consensus on the rules and extent of trading that would be permitted.

These "emissions trading" mechanisms offer the possibility that actual domestic greenhouse gas reductions will be less than the amount required to meet the U.S. commitment. The amount of reductions that can be shifted elsewhere is unclear, but trading offers the possibility of avoiding some of the higher cost/higher impact reductions that otherwise might occur domestically. Indeed, the United States has argued that emission trading is critical to U.S. compliance with Kyoto[24]; a Clinton Administration economic analysis suggested that U.S. compliance costs would drop from $193 per ton with no international emissions trading to $23 per ton with global trading.[25] That the U.S. might shift a sizeable amount of its reductions to overseas sources increases the importance of negotiations to define the trading mechanisms. Defining critical terms, such as "supplemental" and "sinks" were sharply contested issues at The Hague conference, and remain unresolved. With no resolution of these critical variables, any estimate of actual domestic reduction required to comply with the Kyoto Protocol, or of the costs involved, remains problematic.

REDUCTION UNCERTAINTIES

Table 5 depicts the range of reduction requirements implied by the analyses reviewed above. As can be seen from table 5, the projected reductions necessary by 2008-2010 to

---

[24] Statement of Janet Yellen, Chair, President's Council of Economic Advisors, House Committee on Commerce, Subcommittee on Energy and Power, March 4, 1998.

[25] For a discussion of the impact of emissions trading on costs, see: Larry Parker, *Global Climate Change: Lowering Cost Estimates Through Emissions Trading – Some Dynamics and Pitfalls*, CRS Report RL30285, August 20, 1999.

meet the Kyoto Protocol goal vary from over 30% (EIA's high economic growth scenario) to 21% (5-Lab advanced technology efficiency case) below "business as usual" emissions. These high and low projections, which bracket the *Climate Action Report's* point estimate, differ only in assumptions about CO_2 emission: the high projection is 16.7% higher than the low one; they vary from the *Climate Action Report's* by +11.7% to –4.3%.

Table 5. Projected Average Annual 2008-2010 Reduction Requirements to Achieve Kyoto Protocol Emissions Goals

Projection	Projected Greenhouse Gas Emissions in 2010 (MMTCE)	Reduction Required to Meet Kyoto Goal of 1,539 MMTCE	
		MMTCE	%
Base Cases			
Clean Action Report	1,946	407	21
Extrapolation of EPA	2,117	578	27
Composite	2,184	645	30
Economic Cases			
EIA High Growth	2,258	719	32
EIA Low Growth	2,125	487	28
EIA High Elec. Demand	2,240	701	31
Technological Efficiency Cases			
EIA 2001 Tech	2,241	702	31
EIA High Tech	2,109	570	27
5-Lab "Moderate"[a]	2,099	560	27
5-Lab "Advanced"[a]	1,882	343	18

[a] Differences from "Business as Usual" calculated against Composite projection.
Sources: Figures 1-4, Tables 3 and 4.

The reviewed estimates of the reduction necessary to meet the Kyoto Protocol goal range in the base cases from the outdated CAR projection of approximately 400 MMTCE to the composite projection (using EIA's most CO_2 projection) of nearly 650 MMTCE. Varying economic growth assumptions give a potential reduction range from about 490 to over 700 MMTCE. Varying technology and efficiency assumptions give a potential reduction range from under 350 to 700 MMTCE.

Overall, these estimates of the reductions necessary to meet the Kyoto commitment – i.e., the reduction below the "composite" projected emissions for 2008-2010 – range from less than 350 to over 700 MMTCE, with the low projection already assuming "aggressive" policies to reduce emissions. Even ignoring numerous uncertainties, then, there remains a 100% difference from the lower to higher reduction projected necessary to achieve the Kyoto target. To comply with the higher compared to the lower estimate would represent a substantial escalation of effort – and makes estimating potential costs difficult.[26]

None of the estimates integrates the full range of potential variable that could affect CO_2 emissions in the future – economic growth, electricity restructuring, electricity demand, and technological change and penetration. If the lowest projection of emissions were taken for each variable, reduction requirements would shrink more than indicated; conversely, if the highest projection were taken each time, the reduction requirements would be even greater. And for total emissions, none of these estimates explores the uncertainties inherent in the variables affecting projections of the other five greenhouse gases covered by the Kyoto Agreement. In short, projecting future emissions and the amount that emissions might have to be reduced to comply with the Kyoto agreement is fraught with considerable uncertainty.

The precise numerical projections of greenhouse gas emissions or of proposed reductions should be viewed as indicative. They are less accurate than they appear, given the potential for revisions in data and the uncertainties of projections. But in assessing the capability of the United States to comply with the potential commitment to the Kyoto goal, the trend line is inexorably up. None of the reviewed scenarios using assumptions that diminish emissions – low economic growth, putting off retirement of nuclear facilities, accelerated fostering of energy efficient technologies – reverse the upward trend in CO_2 emissions by 2010.[27]

Historical data show that the United States failed to meet its voluntary commitment under the UNFCCC for returning aggregate emissions at the end of the 1990s decade to the 1990 level. Achieving the Kyoto goal would require the continuing upward trend to turn down – indeed, to trend more sharply downward than the rise of the past decade. Even with the potential for emissions trading and sinks to reduce domestic reduction efforts, the need for such a sharp reversal in greenhouse gas emissions trends would represent an extraordinary challenge to U.S. energy and environmental policy.

---

[26] Larry Parker and John Blodgett, *Climate Change: Three Policy Perspectives*, CRS Report 94-816 ENR (1994).

[27] The "advanced" scenario of the "New 5-Labs Study" projects the trend turning downward after 2020.

Chapter 8

GLOBAL CLIMATE CHANGE: LOWERING COST ESTIMATES THROUGH EMISSIONS TRADING – SOME DYNAMICS AND PITFALLS

Larry Parker

OVERVIEW: IMPORTANCE OF EMISSION TRADING TO KYOTO COMPLIANCE

United States concerns about implementing the Kyoto Protocol focus on three interlinked issues: (1) the considerable uncertainty and risk of substantial cost from carbon dioxide (CO_2) abatement; (2) the competitive impacts of compliance, both domestically and internationally, and (3) the comprehensiveness of the Protocol's scope, in particular, the exclusion of third world countries from any CO_2 reduction program.[1] These implementation concerns, along with perceived scientific uncertainty, have prevented any serious effort by the Administration to seek Senate ratification of the Kyoto Protocol.

Removal of any one of the three interlinked issues might significantly improve the prospects for approval of the Kyoto Protocol, or some other regime to control greenhouse gas emissions. For example, if the cost of Kyoto compliance could be shown to be not as burdensome as some have suggested, the competitive impact would be weakened and the concern about comprehensiveness would lessen. Such concerns, along with scientific doubt, would not be eliminated; however, they would be attenuated.

Such a task would not be easy. Estimates of costs to reduce CO_2 emissions vary greatly, and focus attention on an estimator's basic view about the problem and the

---

[1] For a analysis of U.S. policy, see Larry B. Parker and John E. Blodgett, Global Climate Change Policy: From "No Regrets" to S.Res. 98, CRS Report RL30024, January 12, 1999. For a summary of the Kyoto Protocol, see Susan R. Fletcher, *Global Climate Change Treaty: The Kyoto Protocol*, CRS Report 98-2 ENR, updated June 23, 1999.

future, rather than on simple, technical differences, in economic assumptions.[2] Some of these "lenses" through which people view the problem and their effects on cost analysis are summarized in Table1. Based on these perspectives, the cost of complying with Kyoto can appear to range from "none" (or indeed, a positive benefit") to an estimate so high as to potentially bankrupt the economy. For example, the American Petroleum Institute, in summarizing the results of several studies concludes that Kyoto compliance would require: "heavy taxes or high carbon permit prices" to be achieved, resulting in "sharp declines in domestic demand", "encourage imports and reduce exports," and a "significant loss of jobs" on energy related industries.[3] In contrast, a study by a coalition of public interest groups concludes that new energy policies can "cut energy costs, increase employment, and protect the environment." Such a path is seen as reducing energy costs by $530 a household while exceeding the reduction requirements of Kyoto.[4] None of the perspectives on which these analysis are based is inherently more "right" or "correct" than another; rather, they overlap and to varying degrees complement and conflict with each other. People hold to each of the lenses to some degree. The uncertainties about the risk of global climate change and the critical impacts of differing assumptions about the nature of the problem effectively preclude predictions of the ultimate costs of reducing greenhouse gases.

As a result, attention has focused on how to minimize costs by selecting the most economically efficient strategies to reduce CO_2 emissions. With the negotiation of the Kyoto Protocol, the mechanism that has become the centerpiece of this attention is emissions trading. Indeed, Janet Yellen, Chair of the President's Council of Economic Advisors has stated that the "promise of Kyoto can not be achieved without effective emissions trading."[5]

Emissions trading is one of four "flexibility mechanisms contained in the Kyoto Protocol (article 17).[6] Under the Kyoto Protocol, developed countries are given greenhouse gas emissions "budgets" for the compliance period 2008-2012 based on a percentage of their 1990 or 1995 emissions levels (depending on the particular greenhouse gas). If a country determined that it would exceed its emissions limit during the compliance period, emissions trading would permit it to purchase emissions reduction

---

[2] For further discussion, see: Larry Parker and John Blodgett, *Global Climate Change: Three Policy Perspectives*, CRS Report 98-738, August 31, 1998. It identifies three "lenses" through which people can view the global climate change issue, and their influence on cost analysis.

[3] Rayola Dougher, *The Impact of the Kyoto Protocol on Allied Industry Output, Employment and Trade*, American Petroleum Institute, 1999. For a rebuttal of the type of studies on which the API conclusions are based, see Howard Geller, "On Impacts of the Kyoto Protocol on U.S. Energy Markets and Economic Activity." Testimony before the House Science Committee, October 9, 1998.

[4] Alliance to Save Energy, American Council for an Energy-Efficient Economy, Natural Resources Defense Council, Tellus Institute, and Union of Concerned Scientists, *Energy Innovations: A Prosperous Path to a Clean Environment*, Tellus Institute, June 1997. For a rebuttal of the type of studies on which these conclusions are based, see Ronald J. Sutherland, *The Feasibility of "No Cost" Efforts to Reduce Carbon Emissions in the U.S.*, American Petroleum Institute, Issue Analysis #106, May 1999.

[5] Statement of Janet Yellen, Chair, President's Council of Economic Advisors, House Committee on Commerce, Subcommittee on Energy and Power, March 4, 1998.

[6] The other mechanisms are Bubbles (Article 4), Joint Implementation (Article 6), and the Clean Development Mechanism (Article 12).

(i.e., "credits"[7]) from another country that determined it would have achieved more emissions reductions than necessary to comply. With emissions trading, countries that can make relatively inexpensive emissions reductions have an incentive to reduce emissions below the level required by the Kyoto Protocol, and sell the extra credits to other countries whose emissions control costs are more expensive. Thus, both the seller and the buyer would have lower costs by virtue of the seller's profit and the buyer's savings.

This mechanism, however, comes with significant restriction under the Kyoto Protocol. First, emissions trading in restricted to countries that have legally binding greenhouse gas emission limitations. Commonly called Annex 1 parties, only developed, industrialized countries are included.[8] This restriction also applies to two of the other three mechanisms – bubbles and joint implementation projects. Only the Clean Development Mechanism (CDM) can be employed for transactions between Annex 1 countries and countries without legally binding requirements – i.e., developing countries. The specifics of this mechanism are yet to be defined.

A second restriction to trading is the requirement that it "be supplemental to domestic actions for the purpose of meeting quantified emission limitations and reduction commitments..."[9] However, the Protocol is vague as to what "supplemental" means, and the term is subject to continuing negotiation.

Some parties have suggested a third restriction on trading with respect to how reductions are accomplished. Specifically, some have argued that trading be restricted to transactions where the traded carbon credits are the result of explicit controls that reduce greenhouse gases, and not because of economic downturns or other events separate from the Protocol. This issue arises as several countries of the former Soviet Union are projected to have sizeable amounts of credits available for sale because of current economic difficulties. Proponents of trading restrictions argue that such "hot air" reductions would have occurred anyway and would weaken the Protocol's targets. These concerns are heightened by the failure of Russia and seven other members of the former Soviet Union countries to comply with the Montreal Protocol[10]because of "economic difficulties" – the same difficulties that would create the hot air credits.

---

[7] A credit would generally represent the reduction of one metric ton of carbon equivalent emissions.

[8] Although called Annex 1" countries in reference to Annex 1 of the Framework Convention on Climate Change (FCCC), the correct reference is to Annex B of the Kyoto Protocol. The lists of countries in Annex 1 and Annex B are very similar, but not identical. CRS uses the common usage term, Annex 1, in this report.

[9] Article 17, Kyoto Protocol.

[10] The Montreal Protocol is an international environmental treaty designed to protect the stratospheric ozone layer by phasing out the global production of ozone-depleting chemicals, such as chlorofluorocarbons.

However, the Kyoto Protocol places no restriction on the means countries may use to comply with reduction requirements; thus, this position may be difficult to sustain in negotiations.

Because trading is a central feature of costs analyses of the Kyoto Protocol, this paper analyzes some of the dynamics and pitfalls of carbon trading, based on various analyses done on U.S. compliance with the Kyoto Protocol. As the analysis conducted by the Administration has become the focus of much of the cost debate, it receives special emphasis here.

Table 1. Climate Change Perspectives and Policy Parameters

Approach	Seriousness of Problem	Risk in developing mitigation program	Costs
Technology	Is agnostic on the merits of the problem. The focus is on developing new technology that can be justified from multiple criteria, including economic, environmental and social perspectives.	Believes any reduction program should be designed to maximize opportunities for new technology. Risk lies in not developing technology by the appropriate time. Focus on research, development, and demonstration; and on removing barriers to commercialization of new technology.	Viewed from the bottom-up. Tends to see significant energy inefficiencies in the current economic system that currently (or projected) available technologies can eliminate at little or no overall cost to the economy.
Economic	Understands issue in terms of quantifiable cost-benefit analysis. Generally assumes the status quo is the baseline from which costs and benefits are measured. Unquantifiable uncertainty tends to be ignored.	Believes that economic costs should be examined against economic benefits in determining any specific reduction program. Risk lies in imposing costs in excess of benefits. Any chosen reduction goal should be implemented through economic measures such as tradable permits or emission taxes.	Viewed from the top-down. Tends to see a gradual improvement in energy efficiency in the economy, but significant costs (quantified in terms of GDP loss) resulting from global climate change control programs. Typical loss estimates range from 1-2% of GDP.
Ecological	Issue understood in terms of potential threat to basic values, including ecological viability and the well being of future generations. Values reflect ecological and ethical considerations; attempts to convert them into commodities to be bought and sold seen as trivializing the issue.	Rather than economic costs and benefits or technological opportunity, effective protection of the planet's ecosystems should be the primary criteria in determining the specifics of any reduction program. Focus of program should be on altering values and broadening consumer choices.	Views costs from an ethical perspective in terms of the ecological values that climate change threatens. Values such as intergenerational equity should not be considered commodities to be bought and sold. Costs include aesthetic and environmental values that economics cannot readily quantify and monetize.

REVIEW OF ANALYSES: THE DYNAMICS OF TRADING

Several attempts have been made to estimate the cost of U.S. compliance with the terms of the Kyoto Protocol. Seventeen estimates by eight different organizations are shown in Figure 1.[11] In terms of the discussion in Table 1, these estimates are the result of "top-down" analyses, although some have more aggressive assumptions about market penetration rates for new, more energy efficient technologies than others. Several "bottom-up", technology-oriented analyses of potential carbon reductions under various scenarios have been conducted. However, these analyses rely on assumed availability and penetration of various energy-efficient and low carbon technologies, not international emissions trading, to achieve their cost savings, and so are not reviewed here.[12] Indeed, a recently released "bottom-up" analysis conducted by the Tellus Institute expresses concern that the flexibility mechanisms contained in the Kyoto Protocol could threaten environmental integrity and result in misguided policies that could actually increase costs in the long term.[13]

Most analyses of the Kyoto Protocol either exclude emissions trading or limit trading to the developed countries covered by the Protocol (Annex 1 countries). The latter assumption is consistent with the intent and language of the Protocol – developing countries' participation in emission trading is restricted to a "Clean Development Mechanism," the parameters of which are yet to be sorted out.[14] Only three estimates incorporated a global trading scenario. Despite the wide range of estimates plotted in Figure 1 for each trading scenario, the differences between the three trading scenarios strongly suggest that emission trading would significantly reduce the projected costs of U.S. compliance with the Kyoto Protocol. The "promise" of international emissions trading appears to be indisputable, based on existing analyses.

---

[11] The organizations are the Energy Information Administration (EIA), WEFA, Inc., Charles River Associates (CRA), Pacific Northwest National Laboratory (PNNL), Massachusetts Institute of Technology (MIT), Electric Power Research Institute (EPRI), Data Resources Institute (DRI); and the Clinton Administration. A summary of the first seven analyses can be found in Energy Information Administration, *Impacts of the Kyoto Protocol on U.S. Energy Markets and Economic Activity*, prepared for the U.S. House, Committee on Science, U.S. Govt. Print. Office, SR/OIAF/98-03, October 1998, pp. 137-151. Additional scenarios from the CRA analysis are available in Paul M. Bernstein and W. David Montgomery, *How Much Could Kyoto Really Cost? A Reconstruction and Reconciliation of Administration Estimates*, prepared for the American Petroleum Institute, 1998. The Administration's analysis is contained in *The Kyoto Protocol and the President's Policies to Address Climate Change: Administration Economic Analysis*, July 1998.

[12] Indeed, the best known of these studies, the "Five-Lab Study," used carbon taxes of $25 and $50 a ton in developing its scenarios. It should also be noted that the "Five-Lab Study examined technology-oriented strategies to achieve stabilization of U.S. carbon emissions at 1990 levels – not the 7% below 1990 levels required under Kyoto. See: Interlaboratory Working Group on Energy-Efficient and Low-Carbon Technologies, *Scenarios of U.S. Carbon Reductions: Potential Impacts of Energy-Efficient and Low Carbon Technologies by 2010 and Beyond*, September 1997. For a critique of the analysis, see: Energy Information Administration, *Impacts of Kyoto Protocol on U.S. Energy Markets and Economic Activity*, prepared for the House Committee on Science, October 1998, pp. 146-151.

[13] Tellus Institute, *America's Global Warming Solution*, a study prepared for the World Wildlife Fund and Energy Foundation, August 1999, pp. 18-20.

Figure 1 Cost Estimates for Kyoto Compliance (Year 2010)

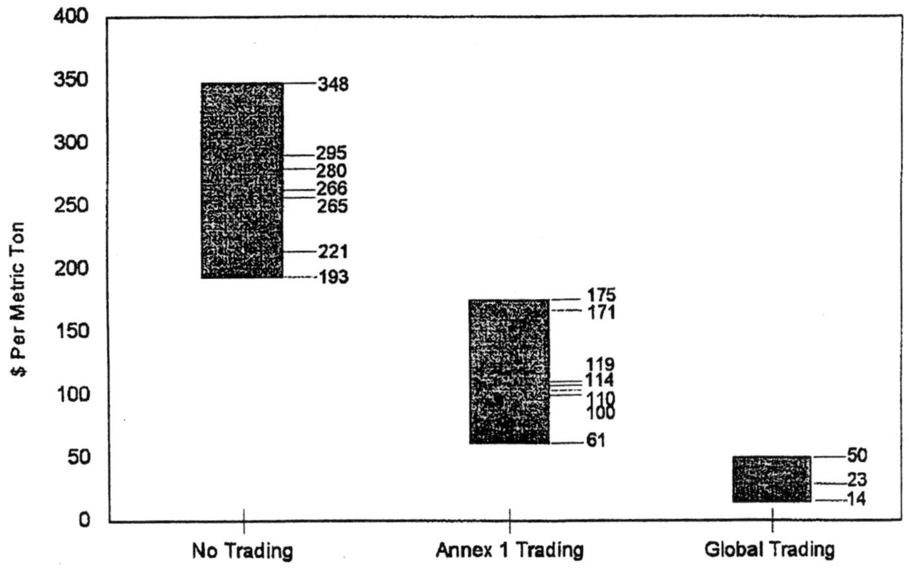

Source: U.S. Energy Information Administration, Charles River Associates, Administration Economic Analysis.

To examine this a little further, two organizations conducted cost analyses for three different trading scenarios. Those estimates, calculated by the Administration and by Charles River Associates (CRA) are provided in Figure 2. As indicated, moving from a no trading posture to an Annex 1 trading posture lowered the cost estimates by 60% (CRA) to 68% (Administration). According to these analyses, if the Kyoto Protocol permitted full global trading, the costs would be lowered by 83% (CRA) to 88% (Administration). These analyses agree on the potential cost reductions presented by emissions trading; a potential that increases as the pool of potential participants increase. This agreement on trading's effect on costs is evident despite the significant disagreement on what the actual compliance costs under the Kyoto Protocol might be.[15]

---

[14] The EPRI analysis included above includes some participation in the CDM.

[15] The difference in costs between the Administration's and CRA's analyses is primarily the result of two factors. First, the Administration assumes a higher energy efficiency improvement rate than CRA. Second, the Administration assumes a higher elasticity of substitution between coal and natural gas. These more aggressive assumptions by the Administration are consistent with its "technological view" of Kyoto implementation. For more on that view, see Table 1.

Figure 2 Comparative Analysis: Administration vs. Charles River Associates

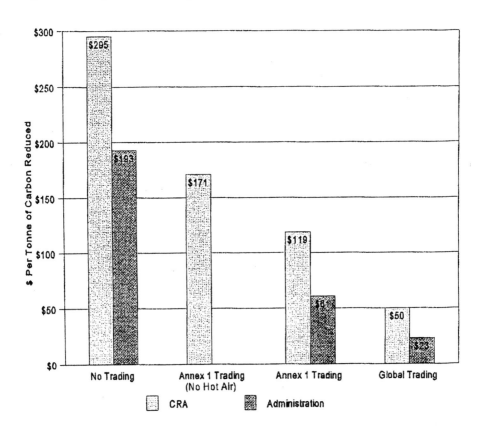

Another important dynamic with respect to trading illustrated in Figure 2 is the importance of "hot air" credits to reducing costs. As noted earlier, "hot air" credits is a rather pejorative term used to describe a potentially large pool of CO_2 credits available from the former Soviet Union. This pool of perhaps 200 million metric tons of carbon, according to DOE estimates, results from the substantial reduction in economic activity in the former Soviet Union since 1990 (the base year for the Kyoto Protocol). If these credits are dumped on the market during the five-year compliance period (2008-2012), credit prices would be depressed, reducing compliance costs as indicated in the CRA analysis. Likewise, without the availability of these credits, the cost of U.S. compliance under Kyoto could be substantially higher.

This substantial cost savings projected under the CRA analysis illustrates why the Administration opposes any restriction on hot air credits. An increase in the available pool of credit sale would tend to reduce the price of credits in the trading market. For a country like the United States, which is projected to be very active in the trading market, lower credit prices would translate into lower compliance costs, all else being equal.

Indeed, the lower cost estimates of the Administration's analysis is partially the result of a trading system that is assumed to be very free and unconstrained. In essence, the Administration assumes the trading system will work very well indeed. Just how well is indicated by Figure 3. In order to gain the 68% cost reduction from emission trading between Annex 1 countries discussed above, 61% of the necessary carbon credits must be bought from other Annex 1 countries. In order to gain the 88% reductions in costs from global trading, 82% of the necessary carbon credits must be bought from other countries. In a maximum trading scenario developed by the Administration, up to 88% of carbon credits would be purchased from other countries.

Figure 3 Administration Analysis: Meeting Kyoto

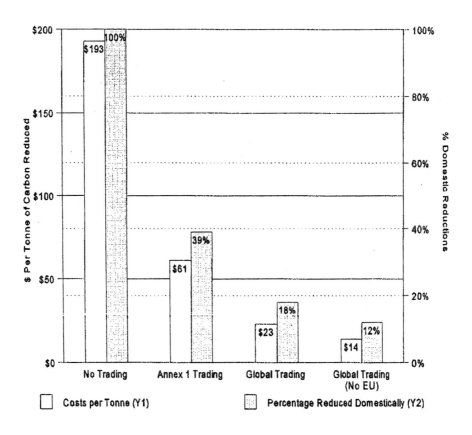

This scale of potential trading may put any resulting, U.S. implementation strategy in conflict with restrictions contained in the Kyoto Protocol, and with the negotiating position of other parties to the Protocol. As noted earlier, according to the Protocol, international emissions trading is to be a "supplemental" implementation tool to domestic efforts. According to the latest European Union position, supplemental means no more

Two conclusions emerge from this review. First, there is little debate among the analyses that emission trading could reduce U.S. compliance cost under Kyoto. Indeed, the percentage reductions resulting from increased trading do not differ greatly. Rather, the dispute is over how well such a program would work. Second, it is the assumption of the Administration's analysis that trading will work extremely well, resulting in substantially lower costs for the United States. This possibility may be difficult to achieve given restrictions contained within the Kyoto Protocol, the negotiating position of some of the other parties, and the sources from which many of the credits are projected to come. Moreover, current efforts to devise a workable trading system suggest that it will be a difficult and lengthy process, at best.

IMPLEMENTATION PITFALLS: COMPARISON WITH ACID RAIN

The importance of trading to cost estimates, and the scope to which it is employed by the Administration in its analysis has no direct parallel in any existing environmental program. The closest example of such a trading program is the acid rain program under title IV of the 1990 Clean Air Act Amendments. However, significant differences between acid rail and possible global warming limit the usefulness of title IV as an analogy for an international carbon trading system. For example, the acid raid program involves up to 3,000 new and existing electric generating facilities that contribute two-thirds of the country's sulfur dioxide (SO_2) and one-third of its nitrogen oxide (NO_x) emissions (the two primary precursors of acid rain). This concentration of sources makes the logistics of emissions trading manageable and enforceable. However, CO_2 emissions are not so concentrated. Although over 95% of the CO_2 generated comes from fossil fuel combustion, only about 33% comes from electricity generation. Transportation accounts for about 20%. Thus, small-dispersed sources in these other areas are far more important in controlling CO_2 emissions than they are in controlling SO_2 emissions. This creates significant administrative and enforcement problems for an international emissions trading program if it attempts to be comprehensive. These concerns multiply as the global nature of the program is considered, along with the number of greenhouse gases that would be included in it.

In addition to the substantive differences in the problems, the trading dynamics of national SO_2 trading and international CO_2 trading are different. As indicated by Figure 4, the largest projected saving from emission trading under the SO_2 program is from permitting relatively simple and uncomplicated trading between a utility's own facilities. An additional ten percent can be gained by permitting intrastate trading. However, expanding the boundaries of the trading to interstate trading does not result in as dramatic cost reductions as for intra-company trading. For implementation policy, this is very significant, as it suggests that the SO_2 trading program does not have to work very efficiently to achieve a large proportion of the economic benefits that have been estimated. Given the increasing regulatory and administrative complexity of expanding the scope of trading to regional levels, the trading dynamics suggest that such complexity can be avoided at little loss of economic efficiency.[6]

Figure 4 Projected Cost of Acid Rain Program

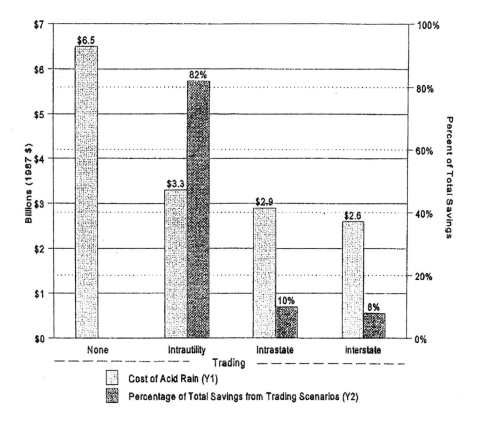

However, although the positive effects of trading have been borne out in the first few years of the SO_2 program, it is not necessarily a harbinger of the potential cost savings from an international carbon trading program. First, the baselines for measuring cost savings are different. For the acid rain program, a unit-by-unit allocation of reduction with absolutely no trading is the baseline from which cost savings from trading are measured. For a carbon trading program, the baseline is an interstate (or intra-country) trading scenario from which cost savings from international trading is measured. Thus, the maximum trading scenario estimated under the acid rain program (interstate trading) is the baseline scenario for measuring the effect of inter-country trading under carbon trading. In essence, the "no trading" scenario of the carbon trading program is the "interstate trading" scenario of the acid rain program. Thus, the scope of international carbon trading is well beyond that of the title IV program.

---

[6] For a further discussion, see: Larry B. Parker, Robert D. Poling, and John L. Moore, "Clean Air Act Allowance Trading," 21 *Environmental Law*, 4, 1991, pp. 2023-2068.

Second, the trading dynamic under the SO_2 program discussed above contrasts strongly with that projected under an international carbon trading program. As indicated in figure 5, under an international carbon trading program, only half the anticipated savings from trading occur in transactions between developed Annex 1 countries. About a fifth of the savings projected by Charles River Associates is the result of "hot air" credits. Finally, about a quarter to a third of the anticipated savings results from transactions with countries not currently covered by the Protocol. Thus about half the total savings from trading comes from sources whose credits are either contested in some quarters, or from countries who are not required to participate in the reduction program at all.[7]

**Figure 5 Savings from Different Carbon
Trading Scenarios: Charles River Associates**

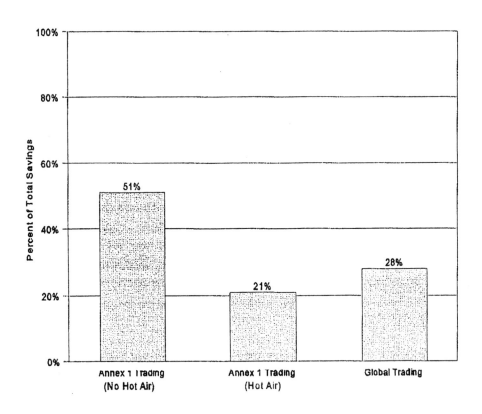

------

[7] Indeed, it is not clear that developing countries can participate in the CDM unless they assume reduction obligations. For more on the CDM, see United Nations Development Programme, *Issues & Options: The Clean Development Mechanism*, United Nations Publications, 1998.

CONCLUSIONS

The potential for international emission trading to reduce U.S. compliance cost under the Kyoto Protocol is substantial and indisputable. Whether that potential can be turned into fact is more problematic. The analysis presented here suggests that implementing international emissions trading under Kyoto would represent uncharted territory for U.S. environmental policy.

First, an international emissions trading scheme has to function very efficiently to achieve the savings projected by analyses. For example, the Administration's analysis relies on an unprecedented amount of international trading to achieve the substantial cost reduction its projects. Under its most aggressive trading scenario, 82%-88% of the U.S. reduction requirement would be bought from foreign sources, resulting in domestic CO_2 reductions of only 66 to 99 million metric tons, compared with an estimated 550 million metric tons if all reduction were achieved domestically.[18] Even restricting trading to Annex 1 countries results in 61% of the country's reduction requirement coming from foreign sources. The magnitude of transactions not only raises questions of its feasibility, but may also conflict with the intent of the Kyoto Protocol – and with the positions of some other countries – which states that international emissions trading is to be "supplemental" to domestic actions.

Second, besides the amount of trading estimated, the sources of these reductions raise additional questions. For example, according to analysis by Charles River Associates, half the estimated savings from international emissions trading would come from either "hot air" credits from the former Soviet Union, or from transactions with Third World countries that are not required to participate in the program. Trading with these sources does not have the certainty that trading with most Annex 1 countries would have in terms of monitoring, enforcement, and integrity of transactions. With respect to the former Soviet Union, current problems with achieving compliance with the Montreal Protocol, a far simpler international treaty, does not bode well for the Kyoto Protocol. That "economic difficulties" are proffered by these countries as grounds for non-compliance and non-enforcement of the Montreal Protocol is particularly disturbing, as those same difficulties are the source of the "hot air" credits.

The situation may be more uncertain with Third World transactions as developing countries generally have neither the incentive of a binding obligation under the Protocol, nor the infrastructure to monitor, enforce, and protect the integrity of transactions. In addition, there may be complications resulting from the Kyoto Protocol itself. As currently written, transactions with Third World countries are to be funneled through a "Clean Development Mechanism," an institution whose role and parameters have yet to be worked out. How much this "middle man" mechanism would affect trades is unclear.

---

[18] For further information on the projected U.S. reduction requirement under the Kyoto Protocol, see: Larry Parker and John Blodgett, *Global Climate Change: Reducing Greenhouse Gases – How Much from What Baseline?* CRS Report 98-235 ENR, March 11, 1998.

Third, the situation with international carbon trading is not analogous to the acid rain program, often cited as a model.[19] The acid rain program involves domestic trading in one pollutant from about 3,000 relatively large stationary sources. As such, it has been administratively manageable, enforceable, and successful. The Kyoto Protocol involves 6 pollutants, million of small, medium, and large sources, and international trading. The maximum trading scenario under the acid rain program – interstate trading – is the baseline for international trading under the Kyoto Protocol. While electric generating facilities – the focus of the acid rain trading program – account for two-thirds of U.S. sulfur dioxide emission, they account for only 29% of the six greenhouse gases emitted in the U.S. (mostly carbon dioxide). Other more dispersed energy uses, such as transportation, make up most of the rest. Add the international scope of carbon trading to this mix, and it is clear that implementation challenges would be on a different level than that encountered with the acid rain program, a level to which the implementation of the acid rain program provides little guidance.

Besides questions raised by the scale of carbon trading, the trading dynamics of carbon trading differ from those of the acid rain program. Unlike an international carbon trading program that must operate very efficiently to achieve much of its cost savings, the acid rain program does not have to do so. The simplest trades – those between a company's own plants – achieve the greatest cost savings under the acid rain program, not interstate trades between unassociated parties. That there have been relatively few interstates trades so far under the acid rain program means once again that the program provides little guidance to any future international carbon trading program.

In short, to expect trading to reduce costs by the 80%-90% suggested by some analyses seems at the current time to be unrealistic. Indeed, the complexity presented by international emissions trading suggest that alternatives may deserve a hearing.

---

[19] For example, see: Council of Economic Advisors, *Economic Report of the President*, U.S. Govt. Print. Off., February 1997, pp. 208-213.

Chapter 9

GLOBAL CLIMATE CHANGE: THREE POLICY PERSPECTIVES

Larry Parker and John Blodgett

INTRODUCTION

Even as the possible role of human activities affecting global climate is being actively questioned, national and international climate change policy actions are being debated.[1] As a signatory to the United Nations Framework Convention on Climate Change, the United States committed to an objective of achieving "stabilization of greenhouse gas concentrations in the atmosphere at a level that would prevent dangerous anthropogenic interference with the climate system"; and to preparing "national action plans" to address emissions of greenhouse gases.[2] The Clinton Administration has seen the possibility of global climate change mitigation as an explicit policy objective influencing the direction of U.S. energy and environmental programs. How proactive that policy should be became subject to debate in 1994 with the release of the Administration's "Climate Change Action Plan."[3]

This debate has been re-ignited by the Kyoto Protocol, agreed to in December 1997.[4] Specifically, under the terms of the Kyoto Protocol, the United States would commit to

---

[1] This paper discusses policy perspectives on the issue and potential actions, but not the underlying controversy concerning the reality and urgency of global climate change – sometimes more narrowly termed "global warming." For background, see Wayne A. Morrissey and John R. Justus *Global Climate Change*, CRS Issue Brief IB89005 [updated regularly].

[2] The Senate consented to ratification of the U.N. Framework Convention on Climate Change on October 7, 1992, with a two-thirds majority division vote; President Bush signed the instrument of ratification of the Convention on October 13, 1992.

[3] William J. Clinton and Albert Gore, Jr., *The Climate Change Action Plan* (October 1993*); Climate Action Report: Submission of the United States of America Under the United Nations Framework Convention on Climate Change* [1994].

[4] On the agreement, see Susan R. Fletcher, *Global Climate Change Treaty: Summary of the Kyoto Protocol*, CRS Report 98-2 ENR, December 22, 1997.

reducing its average annual net carbon-equivalent emissions of 6 gases by 7% below specified baseline years over the 5-year period 2008-2012.[5] If ratified by the Senate, the Kyoto Agreement would move the debate beyond the "study only"[6] and the "no regrets"[7] policies of the Bush Administration, and the present non-coercive, voluntary policies of the Clinton Administration.

The outcome of the Kyoto Agreement is unclear. In July 1997, prior to Kyoto, the Senate agreed by a unanimous vote 95-0 to S. Res. 98, stating that the Administration should not accept an agreement that would seriously harm the economy or that did not require developing countries to meet appropriate reduction requirements. The Administration signed the agreement, saying that costs would not be excessive (particularly because it included emissions trading and joint implementation provisions); and while the Agreement did not impose requirements on developing countries, this issue is to be taken up at meetings in Buenos Aires in November 1998. So the Administration has postponed submitting the Agreement to the Senate for ratification at least until after that time.

Meanwhile, the Administration's current policy takes modest steps to stimulate changes in the technological, economic, and ecological actions of the nation in order to sustain carbon dioxide reduction in the long term. In terms of new technology, the focus of the Clinton plan is on longer term research and development. For example, the Administration's proposed FY1999 budget requested $3.6 million in tax credits and $2.7 billion in new research and development spending over the next 5 years for a new global climate change initiative. The incentives focus primarily on more energy-efficient buildings, industrial cogeneration and control of minor greenhouse gases, fuel-efficient vehicles, and reducing carbon emissions in electricity generation.[8] These actions could be classed as consistent with a "no regrets" approach, the Framework Convention, and U.S. energy policy as articulated in the Energy Policy Act of 1992[9] - although some have

---

[5] On the specifics of the Kyoto reduction requirements, see Larry Parker and John Blodgett, *Global Climate Change: Reducing Greenhouse Gases – How Much from What Baseline?* CRS Report 98-235 ENR, March 11, 1998.

[6] This approach can be summarized as focusing on the study of global climate processes, with particular attention to the potential human role in causing change. It implies taking no action to change human activities on the basis of possible impacts on global climate unless further information verifies the need. This is not to simply "ignore the problem," since it implies focused research with additional resources. Arguably, too, this investment in research would be warranted if a more aggressive plan of action were adopted.

[7] Adopting a "no regrets" policy can be summarized as assessing policy options across the range of federal activities for their potential impact on global climate change, and where alternative policies to achieve a goal otherwise appear similar, adopt the one most consistent with protecting against the risk of global climate change. The idea of "no regrets" derives from the presumption that even if global climate change proves a false alarm, one would not regret adopting policies that are protective if there were no additional (or at most minimal) costs and the policies were justified on other grounds (e.g., have other environmental benefits or energy security benefits).

[8] See Michael M. Simpson, *Global Climate Change: Research and Development Provisions in the President's Climate Change Technology Initiative*, CRS Report 98-408 STM, April 27, 1998; and Salvatore Lazzari, *Global Climate Change: The Energy Tax Incentives in the President's FY1999 Budget*, CRS Report 98-193 E, March 4, 1998.

[9] Besides the numerous titles on energy efficiency and renewable energy, title XVI provides for data collection, analysis, and reporting of greenhouse gases, including a national energy strategy designed to

raised questions about these being "backdoor" efforts to implement the Kyoto Agreement before it is ratified.[10]

In terms of changing market signals to consumers and industry, the Clinton Administration's major concrete policy initiative was the modified BTU tax, originally proposed in 1993.[11] Although not specifically designed to address climate change, the BTU tax would have set a precedent for using energy pricing as a means of effecting changes in consumer preference. Congress rejected the BTU tax proposal, however, and the Administration has not pursued it. Nor did the Administration include a CO2 reduction scheme in its 1998 electric restructuring proposal, S. 2287, as some environmental groups had urged. However, it did make inclusion of emission trading and joint implementation a focal point of its policy position during the Kyoto negotiations, an effort resulting in those mechanisms being included in the final agreements. Also, the Administration's proposed FY1999 budget calls on EPA, assisted by DOE, to analyze options for developing a domestic emission trading system and early reduction program. EPA would work with interested parties to begin building the institutional capacity to implement a tradable permit program.[12]

Because of the enormous uncertainties associated with global climate change – whether global climate change is occurring or will occur, what the effects might be and their magnitude, the consequences that would follow from actions to reduce emissions of greenhouse gases, the costs of actions or of taking no action, the time frame of impacts, etc. – each individual's perception of what, if anything, to do is strongly influenced by personal values, experience, education and training, and outlook in how to cope with uncertainty.[13] These personal variations affect one's definition of the issue and the weight one gives possible approaches to it. This is not just stating the obvious that economists, lawyers, biologists, atmospheric scientists, and others bring different expertise to the issue, or that optimists and pessimists can see the same glass as half full and half empty. This is highlighting the fact that the magnitude of uncertainty accentuates those differences that would apply even if the facts concerning global climate change were indisputable.

---

achieve to the maximum extent practicable and at least cost "the stabilization and eventual reduction in the generation of greenhouse gases."

[10] Wayne A. Morrissey, *Global Climate Change: Congressional Concern About "Back Door" Implementation of the 1997 U.N. Kyoto Protocol*, CRS Report 98-664.

[11] See Lawrence C. Kumins, *The BTU Tax Proposal: House Action, Senate Reaction, and the Transportation Fuels Tax*, Issue Brief IB93061 [archived].

[12] For further discussion, see Larry Parker, *Global Climate Change: Market-Based Strategies to Reduce Greenhouse Gases*, CRS Issue Brief IB97057 [updated regularly].

[13] Implications of differing perceptions are discussed in, for example, John Blodgett *Economic and Environmental Policymaking: Two-Stepping to a Waltz*, CRS Report 94-175 ENR [archived]; Steven Kelman, *What Price Incentives: Economists and the Environment* (Boston: Auburn Publishing Co., 1981); Lester B. Lave and Hadi Dowlatabadi, "Climate Change: The Effects of Personal Beliefs and Scientific Uncertainty," *Environmental Science and Technology*, Vol. 27, no. 10 (1993), 1962-1972; Richard B. Norgaard and Richard B. Howarth, "Climate Rights of Future Generations, Economic Analysis, and the Policy Process," in U.S. Congress, House, Committee on Science, Space, and Technology, *Technologies and Strategies for Addressing Global Climate Change*, Hearings, 17 July 1991 (Washington, DC: U.S. Govt. Print. Off., 1992), pp. 160-173; and "Science and Nonsense in the Global Warming Debate," *ENDS Report* 233 (June 1993), 21-23.

In the end, the origin of and support for different global climate change policy options arise from differing orientations to, or philosophies for, thinking about uncertainty, taking risks, human progress and adaptability, and personal and community values. Differing perspectives of persons affect their observations and interpretations of the issue, influencing their decisions on whether policy interventions are necessary and, if so, what kinds of intervention. At the same time, personal perspectives can change; new knowledge, education, and/or moral suasion may impact on policymaking and individual and corporate behavior, and may also be necessary to create conditions for successfully implementing initiatives relating to climate change.

THREE LENSES FOR VIEWING SOLUTIONS

The many person proclivities and professional constructs that help shape an individual's perspectives on environmental issues in general, and global climate change in particular, can be grouped into three perspectives that affect proposed policies. These perspectives, which can intertwine and overlap, are:

- That environmental problems are the result of inappropriate or misused technologies, and that the solutions to the problems lie in improving or correcting technology
- That environmental problems are the result of market failures, and that the solutions to the problems lie in ensuring that market decisions take into account all costs, including environmental damages; and
- That environmental problems result from a combination of ignorance of, indifference to, and even disregard for, the ecosystem on which human life ultimately depends as well as for the other living creatures that share the planet; and that the solutions to environmental problems lie in developing an understanding of and a respect for that ecosystem and in providing mechanisms for people to express the priority they place on the environment in their daily choices.

Each of these perspectives can be considered a "lens" through which individuals and policy communities view the issue – a lens that provides a particular focus on the nature of the problem and for the kinds of actions to solve it.[14] For shorthand, they might be termed the *technological lens,* the *economic lens,* and the *ecological lens,* respectively.

Each perspective and its associated policy approaches generally are sufficiently distinct that a dominating tendency in policy options can be discerned. As policy frameworks, these lenses incorporate terminology and methods associated with diverse academic disciplines and professions, including not only engineering, economics, and ecological sciences, but also various social sciences, jurisprudence, theology, and others.

---

[14] No further action on global climate change, or setting a policy of no federal government role are options, as well.

As policy frameworks, they should not be confused with any one academic discipline or profession;[15] rather, they are perspectives on policymaking, on how to focus on a policy issue.

While the lenses can be analyzed as distinct perspectives, most of the time for most people they represent predilections rather than conscious alternatives.[16] The lenses differ primarily in what aspects of the issue come into focus, resulting in some being magnified, others obscured, or even distorted. The appropriateness of this focusing is dependent on the characteristics of the specific issue and the orientation of the policymaker. Thus, a policymaker viewing global climate change through one lens – say, the technological lens – is not necessarily disregarding economic or ecological factors, although these factors tend to lie outside, and may be less discernible, than the more clear focus on technological options.

Ultimately, given the diversity of policymakers and the potential overlapping of viewpoints, any global climate policy considered may involve a mix of initiatives representing all of the perspectives. Any mix is likely to reflect mutual accommodation as much as conscious agreement that a combination of approaches better ensures progress toward mitigation goals. The purpose here is not to suggest that one lens is superior to another, but rather to articulate the differing perspectives in order to facilitate communication among different parties and interests.

Technological Lens

Background. Viewed through the technological lens, an environmental problem is an "opportunity" for ingenuity, for a technical "fix." This technologically driven philosophy focuses on research, development, and demonstration of technologies that ameliorate or eliminate the problem. Many uncertainties can be ignored if technology is available to render them irrelevant (a presumption underlying the "pollution prevention" concept, for example). From this perspective, environmental policy entails the development and commercialization of new technologies; Government's role can include basic research, technical support, financial subsidies, economic mechanisms, or the imposition of requirements or standards that stimulate technological development and that create markets for such technologies.

The relationship between environmental protection and technological development was recognized early in the environmental debates and policymaking of the 1960s and 1970s. Particularly in the area of mobile source pollution control, standards anticipated technological development to achieve emissions reductions – commonly called "technology-forcing." Although some in industry argues that this was not an efficient

---

[15] Hence, the economic lens should not be confused with the academic discipline of economics, nor the ecological lens with ecological science. The frameworks are broader than any one discipline, incorporating a range of policy-relevant perspectives, depending on the personal experiences and knowledge of the policymaker.

[16] See Marco Janssen and Bert de Vries, "The Battle of Perspectives: A Multi-Agent Model with Adaptive Responses to Climate Change," *Ecological Economics* 26 (1998), 43-65.

means of encouraging technology (particularly when the deadlines for compliance were short), the process undoubtedly stimulated development.

A "technology-forcing" approach to environmental policy is generally associated with pushing private sector research and development in a socially determined direction (for example, forcing the automobile industry to meet more stringent emission standards than technologically feasible at the time the standards were set). Technology-forcing requirements have also been imposed on public sector programs. For example, the Solid Waste Disposal Act was amended in 1992 to subject Department of Energy (DOE), which is responsible for generating 95 percent of the Nation's mixed waste,[17] to penalties for violating the Act's requirements with respect to handling and disposing of such waste. Because there was no treatment technology available at the time, DOE was required to submit a plan to develop such treatment capacities and technologies to treat all DOE mixed wastes by 1995 (sec. 3021 (b)). Failure to comply was subject to penalties against DOE by EPA.

Regulatory mandates can directly stimulate the commercialization of technology be creating market opportunities. These mandates can be performance-based (meet an emissions level), or technology-based (specify the performance of the technology use). For example, California and four other states mandate that ten percent of the cars sold in those states in 2003 have zero emissions at the tailpipe, this requirement currently can only be met by electric cars. The degree to which these mandates have forced technologies has depended on the perceived seriousness of problems (resulting in accelerated time frames for development, and in very high levels of required performance), the ease of developing the needed technology, and the impact of anticipated costs on consumers.

Along with the use of a regulatory approach to forcing technology, the federal government has also taken an active role in assisting private industry in developing pollution control technology. Some environmentally important industries did not have strong research and development sectors in the late 1960s and 1970s, or did not have ones that would easily be redirected toward pollution control. This led to governmentally directed research and developmental efforts toward pollution control technology. For example, the EPA spent approximately $2 billion supporting development of a feasible flue gas desulfurization (FGD) device for electric utility use to control sulfur oxides. At that time (late 1960s), the utility industry had no central research effort (the Electric Power Research Institute (EPRI) was not started until 1972), and individual utilities devoted their engineering efforts to improving mechanical efficiency of generation, not the chemical engineering necessary for desulfurization. Many utilities also were opposed to adding a chemical process on their plants, preferring other control techniques, such as tall stacks and low sulfur coal. The success of the Government's efforts is indicated by the fact that the FGD device is now the performance and reliability standard by which new, emerging control devices are measured.

---

[17] Mixed waste consists of waste that qualifies as both radioactive (low-level) and hazardous. This means that the radioactive material is subject to the Atomic Energy Act and the hazardous material is subject to the Resource Conservation and Recovery Act (RCRA).

The technological lens reflects a traditional American "can-do" faith in technology, and in the country's ability to find a "technology-fix" to meet the needs of most problems. Such an approach attempts to increase the effectiveness of technology so that social problems can be solved at little or no additional cost. Consumers' desires and needs are taken as a given. The technological response is an effort to achieve an acceptable level of environmental protection without restricting the choices available to those consumers. For example, consumers want to drive. Viewed through the technological lens, policymakers see their role as making that activity less environmentally harmful at minimal cost to consumers, not as restricting that desire or even necessarily as offering alternatives to driving such as mass transit. Efforts to diminish consumer use of the automobile would be seen as a last resort. *The technological lens provides a view of the economy in which technology permits consumers to continue their preferred behaviors while concomitantly achieving environmental goals. It is not necessary for consumers to change their behavior to adjust to the "new reality" of an environmental problem.*

Application to Global Climate Change. Viewed through the technological lens, global climate change is seen as a problem requiring a reorientation of the energy sector from carbon-based fossil fuels to a more "environmentally friendly" energy system based on renewables and conservation. As stated by Worldwatch Institute:

> The end of the fossil fuel gas age is now in sight. As the world lurches from one energy crisis to another, fossil fuel dependence threatens at every turn to derail the global economy or disrupt its environmental support systems. If we are to ensure a healthy and prosperous world for future generations, only a few decades remain to redirect the energy economy.[18]

This view is reflected in a speech of President Clinton on April 21, 1993: the challenge of global climate change "must be a clarion call, not for more bureaucracy or regulation or unnecessary costs, but instead for American ingenuity and creativity, to produce the best and most energy-efficient technology." The focus on technology was evident in the Administration's 1993 Climate Change Action Plan:

> These [long-term] policies must address technologies of energy supply and use, and condition markets for the long-term transition away from activities, fuels, and technologies that generate large emissions of greenhouse gases.

The policies contained in the Action Plan are directed primarily at creating effective markets for investments in existing or nearly commercially available technology that reduce greenhouse gas emissions. The core of a long-term strategy must ensure that a constant stream of improved technology is available and that market conditions are favorable to their adoption. The Action Plan is likely to stimulate a modest acceleration in technological development...Such gains will lay the foundation for the development of

---

[18] Christopher Flavin and Nicholas Lenssen, *Beyond the Petroleum Age: Designing a Solar Economy* (Washington, DC: Worldwatch Institute, December 1990), p. 5.

technologies that could contribute to significant reductions in greenhouse gas emissions in both the United States and abroad...

> Research and development into the technologies that could contribute to greenhouse gas emission reductions will be a critical part of the long-term effort.[19]

These views were reiterated in the President's 1998 $6 billion Climate Change Technology Initiative.[20] As stated by National Economic Council Chair Gene Sperling:

> We think that this [Climate Change Initiative] package is a very good example of what we spoke about when we said that there were win-win opportunities for positive incentives that would clearly show how we can address the issue of climate change and strengthen our economy at the same time.[21]

Looking through the technological lens, policymakers may see technological development as cost-effective, thus improving the economy, not penalizing it. This "win-win" perspective appears clearly in the Administration's 1997 *Climate Action Report*, submitted in accordance with the United Nations Framework Convention on Climate Change.[22] The Plan "combines an array of public-private partnerships to stimulate the deployment of existing energy-efficient technologies and accelerate the introduction of innovative technologies. The goal of these programs is to cut CO_2 emissions, while enhancing productivity domestically and U.S. competitiveness aboard."[23] The cost of a technological approach to the climate change issue appears to net out to zero, or even to save money, depending on how the benefits from increased efficiency are estimated.

The technological lens tends to focus cost-benefit analysis on a "bottom-up" methodology that evaluates the relative costs of projected compliance techniques. As summarized by National Academy of Sciences, "technological costing develops estimates on the basis of a variety of assumptions about the technical aspects, together with estimates – often no more than guesses – of the cost of implementing the required technology."[24] Assumptions are technological, in terms of technological performance; economic, in terms of cost-effectiveness; and behavioral, in terms of penetration rates.

In 1991, the Congressional Office of Technology Assessment (OTA) conducted a "bottom-up" analysis using two CO_2 control scenarios: (1) a moderate scenario focused on available technical options that are cost-effective on a life-cycle basis and seen as presenting no massive problems in terms of market penetration; and (2) a tough scenario focused on the best-available technical options with less concern about difficulties in

---

[19] William J. Clinton and Albert Gore, Jr., *The Climate Change Action Plan* (October 1993), p. 29.

[20] For the status of this program, see Michael M. Simpson, *Global Climate Change: Research and Development Provisions in the President's Climate Change Technology Initiative*, CRS Report 98-408.

[21] As reported in *Daily Environment Report*, "Administration Announces $6.3 Billion Plan of Spending, Tax Credits to Curb Emissions," February 2, 1998, p. AA-1.

[22] Department of State, *Climate Action Report: 1997 Submission of the United States of America Under the United Nations Framework Convention on Climate Change*, Department of State, July 1997.

[23] *Clinton Action Report*, p. 90.

[24] National Academy of Sciences, *Policy Implications of Greenhouse Warming* (Washington, DC: National Academy Press, 1991), p. 48.

market penetration. OTA estimated the moderate scenario would reduce a projected 50% increase in CO_2 emissions from 1987 to 2015 to about 22%. In contrast, OTA estimated that the tough scenario would reduce CO_2 emissions to about half their projected 2015 levels, or 29 percent below their 1987 levels in the year 2015.

OTA estimated that the moderate scenario is achievable at a net savings to the economy; overall fuel savings (such as oil, assumed in the year 2015 to cost about $50 a barrel) would exceed annual operating costs of the control measures. With cost estimates for the tough scenario reflecting more uncertainty about the annualized capital and operating costs of proposed control measures,[25] OTA estimated a range for the tough scenario from a net savings of $22 billion to a net cost of $150 billion annually in the year 2015.

DOE's five National Laboratories – Oak Ridge, Lawrence Berkeley, Argonne, National Renewable Energy, and Pacific Northwest – conducted a more recent effort to estimate the benefits of a technological approach for reducing carbon emission.[26] Called the "five-lab study," the labs analyzed scenarios for technologies to reduce carbon emissions in a cost-effective manner (see table 1). In discussing their results, the National Laboratories concluded: "In short, while there will surely be winners and losers for these energy-efficiency and low-carbon scenarios, our analysis shows that their net economic costs – under a range of assumptions and alternative methods of costs analysis – will be near or below zero."[27]

Table 1: Results of 5-lab Study

Scenario	Direct Costs (billion 1995 $)	Energy Savings (billion 1995$)	Carbon Savings (MtC)
Efficiency Case	$25-$50	$40-$50	100-125
High Efficiency/Low-Carbon Case	$50-$90	$70-$90	310-390

Such a conclusion immediately raises the question: "If technological fixes such as enhanced energy efficiency could actually save money, why aren't people doing to now?" One possible answer is that the projections are wrong: the technological fixes are mirages, and the market has correctly ignored them. An alternative answer, the one focused on by the technology lens, is that widespread commercialization of these technologies is blocked by technological, economic, and institutional barriers. For example, a barrier might be that the initial cost of an energy efficient appliance is higher than a lower efficiency alternative, even though the lifetime cost is less; this can be a

---

[25] OTA estimated the annualized costs of the tough scenario in a range of $350-$570 billion annually. See Congressional Office of Technology Assessment, *Changing by Degrees* (Washington, DC: U.S. Govt. Print. Off., 1991) p. 321.

[26] Interlaboratory Working Group on Energy-Efficient and Low-Carbon Technologies, *Scenarios of U.S. Carbon Reductions: Potential Impacts of Energy-Efficient and Low-Carbon Technologies by 2010 and Beyond,* Prepared for the Department of Energy, 1997.

[27] Ibid., p. 1.17.

barrier to a purchaser who is not aware of the comparative life time costs and/or who cannot afford the upfront cost despite the long-term savings. An activist viewing the problem through the technology lens would look to methods for overcoming that barrier, such as providing information on lifetime costs and/or financial help.

Such proponents tend to look favorably on governmental assistance in overcoming such barriers. This assistance can include public sector research, development, and demonstration efforts; incentives to private enterprise through direct funding, beneficial tax treatment for research expenditures, and cost-sharing programs to help overcome technical barriers and to improve the conditions for commercialization; governmental subsidies to technology; indirect incentives that make existing technologies less attractive than new ones (such as a carbon tax); regulatory interventions that create markets for new technologies; and regulations to address institutional and market barriers, such as energy efficiency labeling requirements.

The technology lens focuses attention on two basic issues: what drives technological development, and what barriers impede it. From this perspective, government can help stimulate the former and help remove the later. For those who envision technological fixes that can achieve environmental goals with minimal economic costs, governmental intervention may be a necessary antidote to market failures and unnecessary barriers. But even for those who would rely primarily on markets and minimize the role of government, the technological perspective is considered optimistic, dynamic, and oriented toward the future.

Economic Lens

Background. Viewing environmental issues through an economic lens focuses attention on markets, price signals, and market imperfections. In this view, the recognition of environmental problems should lead to adjustments in market signals, changing producers' inputs and handling of waster, as well as the composition and level of consumer demand, so as to maximize net social welfare. Cleaning the environment entails costs, which can be weighed against benefits.

The Government's role in this scenario is to ensure the correct market signals. To ensure correct signals, the government can:

- Make consumers and producers aware of information on economic costs and benefits;
- Adjust prices through taxes or fees; and
- Affect supply through tradable permits for products (as with leaded gasoline in the early 1980s) or for production-related emissions (as with sulfur dioxide emissions), or through other market-oriented devices.

With the correct signals, the marketplace can operate to find the optimal solution.

Economic considerations have been an explicit or implicit part of environmental policymaking since environmental quality became a federal issue in the 1960s. The use of economic mechanisms to implement environmental goals was debated in the 1960s and early 1970s, but usually rejected on various grounds.[28]

Excluding economic considerations from environmental protection proved difficult, however. As laws began to be implemented, economic costs became increasingly consequential, although generally masked under "practical" or "feasibility" concerns, as achievement of some environmental standards within specified deadlines proved impossible. Automobile standards were delayed; ozone compliance was postponed; and other issues were litigated. Economic concepts began to re-emerge in the debate over the environment with the need to extend deadlines and to provide more flexibility to polluters to achieve mandated standards.

The preferred economic approach to environmental problems traditionally is the pollution tax. Economists observe that pollution imposes costs on society that are not incorporated in the price of the goods or services responsible for the pollution; these are called "external" costs. An ideal pollution tax "internalizes" these external costs by making the beneficiary of the polluting activity pay for the socially borne costs (polluter pays). As long as polluters find it cost-effective to reduce their emissions to avoid paying the tax, they would add pollution controls until further controls would have higher incremental costs than the tax. When the tax is set at the level at which the marginal costs of more control would equal the marginal benefits society gains by future reductions, society's net welfare is maximized.

Despite the theoretical benefits of the pollution tax methodology, it has received little practical use. Problems of implementation have loomed large, particularly because of a lack of data, especially on benefits. Estimates of the benefits of a specific environmental action can be uncertain and can vary by an order of magnitude. Uncertainties about pollution taxes have focused attention on using economic incentives to increase polluters' flexibility in achieving environmental standards based upon regulation. Unlike a tax that focuses on the price (demand) for a pollutant, these mechanisms focus on the quantity (supply) of the pollutant permitted. During the 1970s, four economic mechanisms were adopted to increase polluters' flexibility in meeting the various requirements of the Clean Air Act. These mechanisms were offsets; bubbling, banking, and netting.

- *Offsets.* The pending failure of many regions to meet the 1977 deadline for the National Ambient Air Quality Standards for ozone presented EPA with the prospect of imposing construction bans on new sources in the con-complying areas. In 1976, EPA proposed to permit new pollution generating facilities that installed Best Available Control Technology to be sited in non-attainment areas if their emissions were offset with reductions from existing sources to that aggregate emissions in the area declined.

- *Bubbling.* In 1979, EPA proposed a bubbling program for existing polluting sources only. A bubbling program permits a facility with multiple pollution sources to treat them as a single source rather than individually. As long as the

---

[28] Steven Kelman, *What Price Incentives: Economists and the Environment* (Boston: Auburn Publishing Co., 1981).

aggregate emissions of the specific pollutant meet the standard, the facility is in compliance.

- *Banking.* In 1979, as an adjunct to their offset and bubbling schemes, EPA promulgated regulations which included banking. Banking permits a facility to save any excess reductions it achieves for future use or sale. In this way, banking facilitates use of offsets and bubbles.

- *Netting.* The oldest (1974) mechanism is also the most successful. Netting permits an existing source under the Act to undergo a major modification without triggering the New Source Performance Standards provided the modification results in no aggregate increase in emissions.

Results from these tradable permit systems are spotty. Netting has clearly been the most used and cost-effective mechanism with estimates of savings of between $0.5 and $12 billion. Other pollutants were added in the 1980s, including the successful lead-trading program. That program indicated the feasibility of a market in "pollution rights" and the value of banking in stimulating trading. EPA estimated that the lead program "saved" refiners over $200 million over the life of the program. This success helped lead to the development of the sulfur dioxide (SO_2) trading scheme contained in the 1990 Clean Air Act.[29]

While this lens is sometimes regarded as the private market's alternative to a regulatory command-and-control program, the interactions are more complex. The so-called "market for pollution rights" would not exist if not for government role in altering what the market would do in lieu of governmental action. If governmental regulations did not restrict sulfur dioxide emissions, there would be no need for sulfur dioxide allowances. Government creates the market and defines the boundaries of acceptable market responses. Under the SO_2 trading program, facilities may buy allowances to meet necessary reductions instead of installing equipment to control pollution. The facilities may not exceed the National Ambient Air Quality Standard for SO_2, regardless of how many allowances their owners hold.

By allowing polluters to choose their lowest cost abatement actions, implementing environmental goals through market mechanisms represents a general elevation of economic "efficiency" as the *sine qua non* of decision-making. Pragmatically achieving this efficiency presumes substantially complete knowledge by producers and consumers of costs, abatement alternatives, and product substitutions as well as substantial flexibility in achieving compliance. The market approach simultaneously maintains the general principle of "polluter pays" as the underlying ethical rationale for the distribution of costs among parties. Through the market, the "polluter who pays" includes not only the producer, but also labor, stockholders, and the consumer (who demands the product and who pays somewhat more for the embedded costs to control pollution).

Those viewing environmental policy through the economic lens generally presume that governmental interference, whether through subsidies or regulation, should be

---

[29] CRS, *Market-Based Environmental Management: Issues in Implementation*, CRS Report 94-213, March 7, 1994, pp. 65-66.

minimal. In reality, the distribution of impacts through the market often leads to calls for political interventions that compromise efficiency and the "polluter pays" principle. The political process tends to weigh relevant differences between various groups affected by an environmental mandate, and special treatment may be deemed necessary to promote justice or fairness. For example, the new sulfur dioxide allowance system contains numerous "special" allocations of allowances to various situations. These special allocations represent subsidies to these groups that a strict "polluter pays" principle would not allow. This is not to say that such allocations are not justified on the grounds of justice or fairness, only that the "polluter-pays" principle is not a distributional principle that policymakers can treat independently of other concerns and criteria.

The economic lens reflects a traditional American belief in individual choice and private markets – given the correct price signals, producers and consumers will adjust their behavior accordingly. This adjustment will be done in the most cost-effective manner, and with a minimum of governmental involvement. Consumers' desires are seen as responsive to price. The issue then is for the price to reflect the costs of relevant externalities. With the right price, supply and demand will find the level that maximizes social welfare. *Policymakers using the economic lens see consumers and producers adjusting their behavior to the "new reality" of an environmental problem by responding to the price signals that take into account a particular environmental goal.* But this approach creates clear winners and losers in terms of who will profit and who will pay the tab. As a result, policymakers adjust governmental intervention to achieve change at a pace and impact that are socially and politically acceptable.

Applications to Global Climate Change. The economic lens focuses policymakers on market-based approaches to address global climate change; these include marketable permit (allowance) programs and various taxes, feed, and rebates,[30] as well as research and development, education, and market-related information. Current proposals for controlling carbon dioxide and other greenhouse gas emissions center on either marketable permits programs (loosely based on the current sulfur dioxide program) or on a carbon tax (the closest analogy is the Chlorofluorocarbon (CFC) tax although there are substantial differences between the two schemes).[31]

Current debate in the United States about implementing carbon reductions has focused on tradable permits. A key element of the Administration's negotiating position at Kyoto was the inclusion of domestic and international emissions trading systems and international joint implementation programs to implement any emission reduction requirements. This support for trading programs has continued after the Kyoto conference. The Administration's FY1999 budget request calls on EPA, assisted by DOE, to analyze options for developing a domestic emission trading system. EPA would work

---

[30] For a general discussion of market-based environmental management, see CRS, *Market-Based Environmental Management: Issues in Implementation*, CRS Report 94-213 ENR, March 7, 1994. For a specific discussion of market-based carbon control, see Larry Parker, *Global Climate Change: Market-Based Strategies to Reduce Greenhouse Gases*, CRS Issue Brief IB97057 [updated regularly].

[31] See *Market-Based Environmental Management: Issues in Implementation*, CRS Report 94-213, pp. 67-70; and *Global Climate Change: Market-Based Strategies to Reduce Greenhouse Gases*, CRS Issue Brief IB97057.

with interested parties to begin building the institutional capacity to implement a tradable permit program. In addition, one bill, S. 687, has been introduced by Senator Jeffords providing for substantial reductions in CO_2 emissions implements through a nationwide tradable permit program.[32]

However, the generally acclaimed success of the sulfur dioxide program at its early stages may not translate easily to a marketable permit program for carbon dioxide. Fundamental differences exist: for example, the acid rain program involves over 2,000 new and existing electric generating facilities that contribute two-thirds of the country's sulfur dioxide and one-third of its nitrogen oxide emissions (the two primary precursors of acid rain). This concentration of sources makes the logistics of allowance trading administratively manageable and enforceable. However, carbon dioxide emission sources are not so concentrated. Although over 95 percent of the CO_2 generated from human activities comes from fossil fuel combustion, only about 33 percent comes from generating electricity. Transportation accounts for about 33 percent, direct residential and commercial use for about 12 percent, and direct industrial use for about 20 percent. Small dispersed sources in transportation, residential/commercial, and the industrial sectors are far more important in controlling CO_2 emissions than they are in controlling SO_2 emissions. This would create significant problems in administering and enforcing a tradable permit program that attempts to be comprehensive or equitable. These concerns multiply as the global nature of the climate change issue is considered, along with other potential greenhouse gases, such as methane and nitrous oxide.

An alternative market-based mechanism to the tradable permit system is carbon taxes – generally conceived as a levy on natural gas, petroleum, and coal according to their carbon content, in the approximate ratio of 0.6 to 0.8 to 1.0, respectively. In the view of most economists, a carbon tax would be the most efficient approach to controlling CO_2 emissions.[33] With the millions of emitters involved in controlling CO_2, the advantages of a tax are self-evident. Imposed on an input bases, administrative burdens such as stack monitoring to determine compliance would be reduced. Also, a carbon tax would have the broad effect across the economy that some feel is necessary to achieve long-term reductions in emissions.

In other ways, a tax system merely changes the forum, rather than the substance of the policy debate. Because paying an emissions tax becomes an alternative to controlling emissions, the debate over the amount of reductions necessarily becomes a debate over the amount of reductions quickly would want a high tax imposed over a short period of time. Those more concerned with the potential economic burden of a carbon tax would want a low tax imposed at a later time with possible exceptions for various events. Taxing emissions basically would remain an implementation strategy; policy determinations such as tax levels would require political/regulatory decisions. One argument for, or against, such a system would be that the tax would raise revenues. The

---

[32] For a further discussion, see *Global Climate Change: Market-Based Strategies to Reduce Greenhouse Gases*, CRS Issue Brief IB97057.

[33] "It is an open and shut case that the most economic way to constrain carbon dioxide (CO2) emissions is a flat-rate tax based on the carbon content of fuels – across the board, no exceptions." David Cope, "Environment, Economics and Science." *UK CEED Bulletin*, No. 53 (Spring 1998), 18.

disposition of these revenues would significantly affect the economic and distributional impacts of the tax.

Other tax schemes to address global climate change are also possible. For example, the European Community has discussed a hybrid carbon tax/energy tax to begin addressing CO_2 emissions. Fifty percent of the tax would be imposed on energy production (including nuclear power) except renewables; fifty percent of the tax would be based on carbon emissions. Another possible approach would be a BTU tax that would focus more on overall energy efficiency and have less impact on the coal industry. A modified version of this approach was embodied in the Administration's 1993 energy tax proposal.

The choice between a tradable permit approach and a tax approach depends in part on one's sensitivity to the uncertainty in the benefits of reductions in greenhouse gases versus the uncertainty in the costs of the program. Those confident of the benefits to be received from reducing greenhouse gases tend to focus on the quantity of pollutants emitted and to argue for a specific, mandated emission level. For example, the Kyoto agreement mandates a specific allowable emission level based on a historical baseline (1990/1995, depending on the gas) for a specific compliance period (2008-2012). While a ceiling is placed on emissions, no ceiling is placed on control costs. Implementing such a reduction program through a market-based scheme, such as a tradable permit program, would probably assure that the costs would be dealt with efficiently through the marketplace; however, those costs are not capped. This is the approach used under the current SO_2 control program. Preliminary results indicate that control costs under the SO_2 program are considerably less than they would have been under an alternative "command and control" scheme. However, there is no lid on the costs, which may rise in the future as the control requirements become more stringent.

Alternatively, a tax in effect places a ceiling on control costs, although the actual reductions achieved are subject to some uncertainty. For example, if a carbon tax of $100 a ton were levied, no polluter would pay more than $100 a ton to reduce carbon emissions. Thus, under worst-case conditions, the program costs would be $100 a ton. However, the actual reductions that such a tax might achieve would have to be estimated, based on economic simulations or actual monitoring. Reductions would not be guaranteed as any polluter could choose to pay the tax rather than to reduce emissions. Reductions could also vary over time as new technology or other events raise or lower the cost of reducing emissions.

A carbon tax or tradable permit program would affect economic behavior in at least three ways: (1) effectively reduce real income through higher prices and therefore reduce overall consumption of goods (particularly in the short-term); (2) encourage manufacturers and consumers to substitute less carbon-intensive (or carbon free) energy sources for current carbon-intensive (i.e., fossil fuel) energy sources; and (3) encourage both research and development of innovative, less carbon intensive or more energy efficient technologies and their penetration into the marketplace. The ability and efficiency of the economy in making these adjustments over a specified period of time would largely determine the impact of a market-induced rise in the costs of energy generated from fossil fuels either through a carbon tax or a marketable permit program.

Depending on the reduction achieved and the model employed, annual gross domestic product (GDP) losses resulting from carbon control are estimated to range from less than one percent to more than four percent, with most falling into a range of one to three percent. If a carbon tax were chosen, that tax would generate revenues – revenues sufficiently large to affect aggregate consumer demand. It is the contractionary pressure of these tax revenues that the Congressional Budget Office (CBO) cites as the major reason for a loss of two percent in U.S. GDP from a $100 per ton carbon tax phased in over 10 years.[34] How the Government chooses to deal with those tax revenues greatly would affect the impact of the carbon tax on the economy. The impact of a carbon tax on the economy would vary depending on a combination of policies beyond just the level of the tax.

The tax level necessary to achieve a given reduction is also subject to a wide range of estimates. The Stanford Energy Modeling Forum compared 13 models under a series of control scenarios with common assumptions (where possible), including one calling for carbon emissions stabilization at 1990 levels by the year 2000.[35] About half of the models studied estimated the carbon tax necessary to meet the stabilization target in the year 2000 to be about $30 per ton or less, while the other half estimated the necessary carbon tax to be about $100 or more.

With respect to a tradable permit program, the Administration testified in March 1998, that a carbon trading program among developed countries could reduce U.S. compliance costs under the Kyoto Agreement by an estimated 60-75% compared with a compliance strategy that allowed no trading. Full participation by developing countries in a trading program is estimated by the Administration to reduce U.S. compliance costs by an additional 55%.[36] Details on this analysis were released in July. Included in those details was an estimate that permit prices under Kyoto would be in the range of $14 to $23 per ton of carbon equivalent.[37]

In response, the American Petroleum Institute funded a study to determine the assumptions underlying the Administration's conclusions.[38] The study, conducted by Charles River Associates (CRA), agreed with the Administration that full global trading, if achievable, would significantly reduce costs. CRA estimates a worst-case scenario (i.e., no trading at all) at $295 per metric ton in the year 2010. If trading is achievable only among developed countries, the international permit price would be about $120 per metric ton, if excess permits are offered by Russia ($171 per metric ton if not). If global trading is achievable, the permit price would be about $50 per ton. Using a different

[34] Congressional Budget Office, *Carbon Charges as a Response to Global Warming: The Effects of Taxing Fossil Fuels* (August 1990), pp. 35-37.

[35] Energy Information Administration, *Energy Modeling Forum Study 12 – Global Climate Change: Energy Sector Impact of Greenhouse Gas Control Strategies*. Response to request by the House Committee on Energy and Commerce (May 4, 1992).

[36] Council of Economic Advisors, Testimony of Dr. Janet Yellen, "The Economics of the Kyoto Protocol," Hearings before the Senate Committee on Agriculture, Nutrition, and Forestry, March 5, 1998.

[37] White House, *The Kyoto Protocol and the President's Policies to Address Climate Change: Administration Economic Analysis* (July 1998).

[38] Paul M. Bernstein and W. David Montgomery, "How Much Could Kyoto Really Cost? A Reconstruction and Reconciliation of Administration Estimates," Charles River Associates, June 1998.

model, CRA reconstructs the Administration's analysis which suggest a no trading cost of $193 per ton, a developed country trading scenario cost of $23 per ton, and a global trading scenario cost of $14 per ton – in line with the estimates released by the Administration. CRA states that the Administration analysis is internally consistent and compatible with mainstream economic analysis, but makes very optimistic assumptions about the economy's ability to reduce emissions at low cost and the potential flexibility the country would have to purchase permits internationally.[39]

Because the problem of greenhouse gas emissions is seen in terms of internalizing a currently external cost, the economic lens implies that the marketplace can solve the problem if given sufficient incentive with minimal governmental interference. The Government's role primarily consists of providing a market-based signal to private industry about the external cost (e.g., emission taxes, tradable permits, etc.). In reality, the Government's role is more involved. For taxes, this includes determining its level, any phasing-in period, escalation, and recycling of revenues received. For permits, this includes the total numbers of permits allowed, initial allocation formulas, any phasing in period, penalties, transaction procedures, and tax liability. While an economic approach would supplement the policy process in implementing a greenhouse gas reduction program, it would not be a substitute for basic policy decisions and oversight.

A limited or supporting governmental role is consistent with the overall perspective of the economic lens: private initiative, economic cost-effectiveness, concern about impact of environmental policy on economic policy, cost aversion, and reliance on market force.

Ecological Approach

Background. The development of environmental protection as a national policy concern reflects three factors: (1) the development of an environmental consciousness among the electorate, (2) a change in the climate of decision-making among individuals, businesses, and government at all levels, (3) the availability of opportunities to make concrete decisions based on environmental grounds (either in addition to or in opposition to economic criteria).

The underlying basis of an environmental consciousness is an understanding of the interconnectedness of the planet's biological processes, and a recognition that changes caused by humans may have ecological effects beyond those intended or foreseen. From this perspective, it is in humanity's self-interest (as well as in the interests of non-human life) to protect the basic biological processes that are the foundation of all life; humans can protect those processes by being conscious of humanity's environmental impact and by avoiding or mitigating that impact to the greatest extent necessary (accepting that some impact is unavoidable, and that ecological science has a crucial role in discovering the effects of human activities).

---

[39] Charles River Associates, p. 18.

A seminal characterization of the ecological perspective is *A Sand County Almanac*, by Aldo Leopold.[40] He suggests that humankind has developed two ethical dimensions – the first dealing with the relation between individuals and the second with the relation between the individual and society. But, says Leopold:

> There is as yet no ethic dealing with man's relation to land and to the animals and plants which grow upon it...The extension of ethics to this third element in human environment is, if I read the evidence correctly, an evolutionary possibility and an ecological necessity.[41]

Describing the need for an "ecological conscience," Leopold concluded that the environmental problem "is one of attitudes and implements"; the development of a "land ethic" requires "an internal change in our intellectual emphasis, loyalties, affections, and convictions."[42]

The challenge of the ecological approach was given global scope by the "Brundtland Report" of the World Commission on Environment and Development. Articulating the goal of "sustainable development," its forward describes the challenge this way:

> If we do not succeed in putting our message of urgency through to today's parents and decision makers, we risk undermining our children's fundamental right to a healthy, life-enhancing environment. Unless we are able to translate our words into a language that can reach the minds and hearts of people young and old, we shall not be able to undertake the extensive social changes needed to correct the course of development.
>
> ...We call for a common endeavor and for new forms of behavior at all levels and in the interests of all. The changes in attitudes, in social values, and in aspirations that the report urges will depend on vast campaigns of education, debate, and public participation.[43]

The idea of "sustainable development" suggests future generations should enjoy the same opportunities for meaningful and fulfilling lives as the current generation. A sustainable society has been defined as "one that satisfies its needs without jeopardizing the prospects of future generations."[44] The concept thus serves as an umbrella to encourage development of renewable resources and conservation on non-renewable resources.[45]

---

[40] Aldo Leopold, *A Sand County Almanac, with Essays on Conservation from Round River* (New York: Ballantine Books, 1970), pp. 237-264.

[41] Ibid., p. 239.

[42] Ibid., pp. 263, 246. Some, viewing global climate change through the ecological lens, see in the long-term risks an indictment of the lifestyle and economic structure of Western society – a viewpoint profoundly disturbing to others who do not look through the same lens. As noted in Leopold, an environmental ethic imposes new obligations, calls for sacrifice, and changes existing values.

[43] *Our Common Future* (New York: Oxford University Press, 1987), p. xiv.

[44] Lester R. Brown, et al. *State of the World, 1990* (New York: W.W. Norton & Company, 1990), p. 171.

[45] See, for example, Richard B. Norgaard and Richard B. Howarth, "Climate Rights of Future Generations, Economic Analysis, and the Policy Process," in U. S. Congress, House, Committee on Science, Space, and Technology, *Technologies and Strategies for Addressing Global Climate Change*, Hearings, 17 July 1991 (Washington, DC: U.S. Govt. Print. Off., (1992), pp. 160-173.

The emergence of the ecological perspective (or the "land ethic" or "sustainable development") is manifest in new values and practices of individuals, businesses, and Government.

Within the Federal Government, the National Environmental Policy Act of 1969 represented a watershed in establishing the principle that major federal decisions should publicly disclose and take into account environmental impacts. Originally resisted by many agencies, the idea of assessing the environmental consequences of decisions through "Environmental Impact Statements" has now become routine. Also, over the past two decades, the Federal Government also has taken steps to foster public awareness of environmental values through support for environmental education. In addition, the Federal Government has used procurement policies to support environmental goals; for example, by requiring purchases of paper of minimal recycling content and authorizing payment of a premium for it, and has revised statutes to make Federal facilities subject to these requirements.

The change in societal values resulting from an increased ecological consciousness also affects the perspectives of corporate decision-makers. Despite the often-confrontational relationship between federal environmental policymakers and industry, a consequence often attributable to the command-and-control regulatory approach to environmental policy, industry itself has increasingly recognized that community environmental values are part of the social milieu in which industrial production occurs.

A 1994 article in the chemical industry publication *Chemical Week* reviewed the industry's perceptions of pollution control. It noted that, in the early 1970s, most corporations viewed environmental management as a "threat" and that pollution control expenditures were "non-recoverable investments."[46] The article observed that, in 1970, "economist Milton Friedman described the actions of any company making pollution control expenditures beyond that 'required by law in order to contribute to the social objective of improving the environment' as 'pure and unadulterated socialism'." In contrast, today, major corporations are espousing the benefits of proactive environmental management, stewardship, and environmental leadership. The chemical industry, which was suffering from poor public perceptions, particularly after the Bhopal incident, has been at the forefront of this shift, as indicated by remarks of Robert Luft, Senior Vice President of Du Pont Chemicals: "Our continued existence requires that we *excel* in safety and environmental performance...We must shift our mindset from 'meeting regulations' to 'meeting public expectations'."[47]

This new attitude, or climate, of decision-making is providing many businesses and individuals with new alternatives and opportunities to choose environmentally preferred options wither in concert with more traditionally based economic criteria or in opposition of such "self-interest"-based criteria. For example, the chemical industry today sponsors

---

[46] "34 Years of Environmental Strategy," *Chemical Week* (August 24, 1994), 27.

[47] Robert v.d. Luft, "Protecting the Environment: It's Good Business," Remarks, at the National Petroleum Refiners Association International Conference, San Antonio, Texas (26 March 1991), p. 9. For recent views of chemical industry corporate managers, see "Responsible Care – An Industry Takes Its Bearings," *Chemical Week* (July 1/8, 1998), 33-128.

a "Responsible Care" campaign,[48] and prodded by environmental groups and EPA, the Chemical Manufacturers Association has committed the industry to testing of high-use chemicals.[49] An independent but related initiative is the Green Chemistry Institute, a nonprofit organization with the mission of promoting pollution prevention using "economically sustainable clean production technologies."[50] In addition, EPA and the American Chemical Society jointly sponsor annual "Green Chemistry Challenge Awards" to recognize pollution prevention through innovative chemistry; the first Green Chemistry Award was presented in 1996.

Individuals, as consumers and citizens, are also exercising options to express and environmental consciousness that extends beyond immediate economic self-interest. Consumers' responses to such environmental problems can and will reflect environmental values. For example, recycling programs have increased in recent years, despite questionable economics and the significant consumer inconveniences involved. Such a trend suggests the power of aesthetics and the perceived intrinsic value of the environment as a force which influences people's preferences and priorities.

The ecological lens magnifies elements that are psychological, philosophical, and theological.[51] A policy decision to address a pollution problem generally involves a sophisticated and sometimes lengthy educational process of which economics and technological availability are only components. In this view, environmental education, Smokey the Bear, and environmental interest groups from Audubon to Greenpeace to Zero Population Growth represent efforts to inculcate the sense of moral obligation toward the environment – to acculturate people to the importance of the environment as essential to long-term human health and welfare. Such efforts can promote a climate of opinion in which environmentally responsible decisions are socially endorsed and environmentally irresponsible decisions are stigmatized as not socially acceptable. Pollution protection gets on the national agenda not on the basis of affordability or whether control technology exists, but because an environmental problem is recognized as a threat to human health or welfare. *The ecological approach understands the problem of environmental policy implementation to be the moral education of individuals and institutions of the dimensions of the ecological crisis, changing the climate in which decisions are made, and providing opportunities for individuals and institutions to make decisions based on ecological concerns, rather than having those choices limited to alternatives dictated solely by economic criteria.*

Application to Global Climate Change. In some ways, global climate change is the quintessential issue for an ecological lens, as it so clearly involves far-reaching dimensions including the standing of future generations, non-human life, and distributional justice around the globe. The ecological lens provides a decision criterion in the face of uncertainty or of competing preferences. Aldo Leopold observed that the land ethic "may be regarded as a mode of guidance for meeting ecological situations so

---

[48] "Responsible Care – A Revolution Hits the 10-Year Mark," *Chemical Week* (July 1/8, 1998), 33-129.

[49] Peter Fairley, "CMA confirms Extensive Testing Gaps for Major Chemical," *Chemical Week* (June 24, 1998), 9.

[50] Bette Hileman, "Virtually Green," *Chemical & Engineering News* (June 8, 1998), 31-32.

[51] Leopold noted that Ezekiel and Isaiah decried the despoliation of the land.

new or intricate, or involving such deferred reactions, that the path of social expediency is not discernible to the average individual."[52] No situation is better described as "no new and intricate" or as having "such deferred reactions" than global climate change.

An ecological perspective on global climate change focuses attention on an enlightened public to implement stewardship through a changed value system. Numerous international and domestic entities are supporting activities to foster governmental, corporate, and public awareness of the global climate change issue and to encourage remedial actions. (Other entities provide "neutral" information and analysis on the issue, and sill others actively lobby against the viewpoint that action is justified at this time.[53]) These organizations support activities that translate into concrete actions through a variety of mechanisms, including voluntary programs for businesses and alternative "green" options that allow for individual consumers to make ecologically responsible decisions even when they cost more than do traditional choices.

The current umbrella for activities to foster action in the U.N. Framework Convention on Global Climate Change, under which a range of activities, from research and development to education, are sponsored. Manifesting the ecological perspective, the Framework Convention defines the signatories' objective to be the protection of ecosystems from "dangerous anthropogenic interference with the climate system...to allow ecosystems to adapt naturally to climate change, to ensure that food production is not threatened and to enable economic development to proceed in a sustainable manner."[54] Economic and human concerns are seen as interdependent with ecological processes. The potential policy agenda could include virtually all human endeavors and relationships, from industrial policy to North-South restructuring of international institutions. The Kyoto Protocol, completed in December 1997, is indicative of the breadth of effects that controlling global climate change may entail.

From the ecological perspective, achieving such a broad policy agenda would require an active federal governmental role that involves educating the citizenry about the need to act on the risk of global climate change, providing the public with a role model in terms of government's own decisions and priorities, and developing opportunities for individuals to make ecologically responsible decisions even if those decisions are not economic in a traditional sense. At this stage of the climate change debate, the federal role has included four kinds of activities that reflect environmental stewardship.

- First, making decisions that take into account potential consequences for global climate change and taking actions that support and promote environmentally "friendly" products and processes (for example, through procurement policies or through product labeling).
- Second, internationally exploring the possibilities of achieving consensus on further reductions and on inter-related economic and human issues.

---

[52] Leopold, p. 239.
[53] For links to diverse efforts to inform the public about global climate change, see the CRS Global Climate Change Briefing Book, at: [http://Thomas.loc.gov/brbk/html/ebgcclnk.html].
[54] United Nations Framework Convention on Climate Change, article 2.

- Third, supporting education of the public on environmental concerns generally and about global climate change specifically, and fostering the inculcation of environmental values in educational programs.
- Fourth, developing mechanisms that permit the public to express their environmental values in everyday decision-making.

Similar activities are being promoted through various corporate and nonprofit initiatives, as well. For example, a 1998 corporate initiative under the auspices of The Pew Center On Global Climate Change[55] is designed to bring "a new cooperative debate on climate change." Accepting "the views of scientists that enough is known about the science and environmental impacts of climate change for us to take actions to address its consequences," the Center believes "business can and should take concrete steps now in the U.S. and abroad to assess their opportunities for emission reductions, establish and meet their emission reduction objectives, and invest in new, more efficient products, practices and technologies." Besides this commitment to stewardship, "major companies and other organizations are working together through the Center to educate the public on the risks, challenges and solutions to climate change"; undertaking "studies and policy analyses that will add new facts and perspectives to the climate change debate in key areas such as economic and environmental impacts, and equity issues"; and engaging in an international effort designed to increase the global understanding of market mechanisms, and to work with developing countries to assess emission reduction opportunities."

The ecological perspective emerges from individual actions both in terms of support for educational endeavors – as in support for environmental interest groups – as well as through market choices based on ecological impacts rather than on pure economic costs. Indeed, these actions can go against prevailing economic or technological trends. For example, people may choose to pay more for a product or a service because it is perceived as being more "green" or "climate friendly" than alternatives based on traditional economic or technological considerations. In a sense, customer preferences can outrun the marketplace by creating a demand for a product that producers did not anticipate. In such cases, economic and technological mechanisms follow the ecological imperative, rather than defining limits to it.

An example of such a mechanism is "green pricing" currently being introduced in California.[56] Developed within the context of the continuing restructuring of the electric utility industry, green pricing permits consumers to choose their electricity supplies according to the environmental impact of their generation. Because some renewable resources and other "green" sources of electricity currently cost more to produce than

---

[55] The efforts are spearheaded by the Center's Business Environmental Leadership Council whose members include: American Electric Power, Boeing, Company, BP America, Enron Corp., Intercontinental Energy Corporation, Lockheed, Maytag, The Sun Company, 3M, Toyota, United Technologies, U.S. Generating, Whirlpool Corporation. The quotations in this paragraph are from the Pew Center on Global Climate Change's website: [http://www. pewclimate.org/home.html].

[56] For a general discussion of green pricing, see Steve Pickle and Ryan Wiser, "Green Power Marketing," *Public Utilities Fortnightly*, December 1977, pp. 30-35.

more conventional sources, consumers would have to pay more for it. However, early indications from pilot programs in New England indicate a potential market for such power, particularly among residential consumers.[57] This indication is also coming from California where several companies are providing consumers with the option of cleaner electricity than the current mix of generation.[58] These options come at some additional costs for consumers, but it appear that for some consumers, at least, environmental considerations may compete with cost in deciding on electricity suppliers. Time will tell how many consumers will trade off costs for environmental values, and how large a premium they will be willing to pay – but in any event provision of such alternatives requires governmental initiatives.[59]

Many actions to reduce emissions of greenhouse gases serve multiple social ends – such as energy conservation and pollution prevention that improve the efficiency with which human needs are met. Governments and corporations have taken a lead in fostering energy conservation and efficiency in use, particularly in developed countries. In the U.S., EPA and DOE sponsor a range of energy efficiency programs, including "Green Lights," "Energy Star Buildings," Energy Star Products," "Climate Wise," "Consumer Labeling," and others. These programs include public-private partnerships that promote energy-efficient lighting, buildings, and office equipment.[60] DOE funds research and demonstration, pursuing energy efficiency in transportation, industry, utility, and buildings sectors.[61] There is also an Alliance to Save Energy, a nonprofit coalition of prominent business, government, environmental, and consumer leaders who promote the efficient and clean use of energy worldwide, arguing benefits for the environment, the economy, and national security.[62]

These EPA and DOE activities fall within the Administration's Climate Change Technology Initiative. While technological in thrust, a key element of many of these programs involves education of prospective consumers to persuade them not only of potential cost savings but also of social benefits to be gained. Thus technology (and markets) can be the tool for effectuating the moral imperative driven by the ecological perspective.[63]

Similarly, government and corporate initiatives for pollution prevention, through, for example, source reduction and product stewardship, foster systematic changes that have the potential to reduce global climate change risks.

With a public more aware of the problem of global climate change and with the availability of relevant technological and/or economic alternatives, the implementation of

---

[57] Ibid., p. 31.
[58] For a description of alternatives available in California, see the Natural Resources Defense Council website at [http://www.igc.org/nrdc/howto/encagp.html#results].
[59] Alexi Clarke, "Buyer Beware," *New Scientists* (13 June 1998), 49.
[60] See, for example, the White House Initiative on Global Climate Change, "Fact Sheet on Potential Industry Sector Savings" (October 22, 1997), at [http://www.epa.gov/oppeoee1/globalwarming/actions/global /clinton/savings.html].
[61] Fred Sissine, *Energy Efficiency: The Road to Sustainable Energy Use?* CRS Issue Brief IB97027.
[62] [http://www.ase.org].
[63] However, some "deep ecologists" reject technological fixes and the use of market mechanisms on the grounds that they merely further a non-sustainable system that needs to be replaced.

the broader agenda through appropriate measures becomes possible: making available options that permit people to exercise their moral obligation.

THE THREE LENSES AND POLICY APPROACHES

Each of the three lenses implies fundamentally different ways of assessing policy actions to address global climate change. Crucial variations emerge in perspectives on cost analysis, scientific uncertainty, and the role of government.

Cost Analysis as Viewed through the Lenses

The technological lens focuses attention on the outcome of the innovation; actions are justified if they resolve the pollution problem, and costs and benefits should be weighed in terms of the outcome, not in terms of the transitional costs. In contrast, those viewing the issue through the economic lens tend to focus on costs and benefits as the critical metric for evaluating policies; actions are justified when the benefits outweigh the costs, but not otherwise. The ecological perspective basically suggests that policy choices can be based on a recognition of "rights" rather than costs and benefits, the principles of protecting life and of preserving the ecosystem for future generations govern choice.

These differing viewpoints have implications for the timing and focus of invested resources. Looking through the technological lens, a policymaker would focus on investing resources directly in technical options. Some investment in understanding the problem may be necessary to delineate technical options, but new technologies may make extensive research in understanding the problem moot (as when a process change eliminates use of a chemical of concern). Looking through the economic lens, a policymaker would typically first invest resources in understanding the problem and the costs and benefits of alternatives. That assessment would reveal whether society would be better off adopting policies and committing resources to action – e.g., to reduce carbon dioxide emissions. Looking through the ecological lens, a policymaker who perceives a risk to health and/or ecological systems would tend to promote immediate action. Investments in understanding the problem and the costs and benefits would be undertaken only to the extent appropriate to ensure cost-effectiveness of those actions. Because the ecological lens portrays benefits largely in non-economic terms (sustainability, equity), efforts to quantify and monetize those benefits may be viewed as inappropriate – even immoral. Instead, people are provided with alternatives to act on the problem, allowing them to choose a "responsible" option, even if it costs more than a traditionally defined "economic" option.

Technological Lens. Those using the technological lens see it as a "far-sighted," economically justifiable approach to global climate change. Technology is seen as the impetus for improved efficiency in the economy, concomitant with improved environmental protection. Although the development of technology may be encouraged for a variety of reasons, its commercialization if ultimately based on cost-effectiveness.

In terms of the substance of the environmental issue, the user of the technological lens is typically agnostic or indifferent. The current economic system is viewed as inefficient when economic decision-making is considered on a "life-cycle" basis. When considered on this broader perspective, reductions in carbon emissions may be possible at no net costs to the economy – even at net savings.

Under the technological lens, the parameters of cost analysis change. Concepts like "life-cycle" costs are pivotal in making the cost-effectiveness case for new technology. Existing barriers (institutional or financial) to the rapid and widespread commercialization of new technologies are seen as artificial constraints to be overcome by government and individuals. The focus of analysis is on cost-effectiveness of solutions, not so much on the benefits of the policy.

Economic Lens. The view through the economic lens fits the global climate change issue within the boundary of market economics. The motivations of people in reducing pollution is unimportant; the critical assumption is that people will act in their own self-interest as dictated by price signals. The global climate change issue becomes another consideration in setting prices – an externality than needs to be internalized. If that price increment does not result in significant reductions, it is because none is economically justified.

Under the economic lens, the potential impacts of controlling greenhouse gases on the economy versus expected benefits is a central variable in determining the degree and time frame of reductions. Economic efficiency is the primary criterion for assessing emission reduction programs. Any existing inefficiencies in the economic system are assumed to reflect market reality and to be difficult to eliminate (and eliminating them may be undesirable). Uncertainty about the potential benefits is understood to be a factor in determining the stringency of any reduction program and a potential reason for stretching out compliance. For this lens, cost-benefit analysis is very important in assessing potential control programs. To the extent that new technology is projected to be cost-effective and can overcome any existing market barriers or distortions, its impacts will be included in the cost part of the cost-benefit equation.

Ecological Lens. Those looking through the ecological lens are suspicious of attempts to measure the economic effects of global climate change options. Most efforts to measure economic effect involve comparing a carbon control scenario with a "baseline" projection. The baseline generally is defined as the path the economy would take assuming no changes attributable to adoption of climate change policies.

However, the baseline also tends to connote a path with no distortion; it is the path from which distortions are measured. This conveys some normative legitimacy on the baseline. If global climate change arguments are correct, then the current path is not sustainable in the long run, and the baseline means little – a concern reflected in on-going efforts to incorporate "green accounting" into major economic indicators, such as the Gross Domestic Produce (GDP).[64] Arguably, if an ecological perspective returned the actual path to long-term sustainability, that scenario would represent the more reasonable

---

[64] Carol S. Caron, "integrated Economic and Environmental Satellite Accounts," *Survey of Current Business* (April 1994), 33-49.

baseline. Discussions of economic "growth" and "distortions" are relative to one's perspective on the long-term potential for economic growth in a world with increasing carbon dioxide concentrations.

Commonly, those looking through the ecological lens tend to dismiss economic cost analysis, and particularly cost-benefit analysis, as being of limited usefulness in the overall debate on global climate change, while acknowledging that they can have utility in developing and choosing specific options. From the ecological perspective, people should respond to the global climate change crisis because of its threat to important values, such as the fate of future generations, not because action can be justified on the basis of some narrowly defined cost-benefit analysis. Traditionally, such analysis tends to place value only on those benefits that can be easily quantified, while discounting or ignoring many values that would be seen as governing through the ecological lens. Viewed through the ecological lens, lives and such values as intergenerational equity should not be quantified as a commodity.[65] In this view, treating the fate of future generations in terms of cost-benefit analysis and market forces should be accorded the same social condemnation allotted those who "prostitute" themselves by selling something for money that should not be sold. What people need are alternatives to many of the choices that the marketplace provides based on traditionally defined economic considerations.[66]

At the same time, a burgeoning area of study is ecological economics, and in particular analyses to determine the economic benefits of ecosystems services, which include climate regulation.[67] Such studies may serve to defend environmental values that are rarely accounted for in traditional economic analyses; they also provide another example of the intertwining of the viewpoints.

The Role of Science as Viewed through the Lenses

Although some would prefer that science dictate the timing and magnitude of environmental policymaking, the nature of environmental science (and environmental policymaking) is not such that definitive guidelines are likely in any significant issue. Scientific knowledge represents a continuum of knowledge and uncertainty; policy initiatives go forward when a sufficient majority of the society concludes that what is known about the problem outweighs the uncertainties, or that the risks of delay despite uncertainty are not acceptable. In some cases, increases in knowledge about an

---

[65] The ecological view was shown in the negative response to an economic analysis prepared for the U.N.'s Intergovernmental Panel on Climate Change; "The Social Costs of Climate Change: Greenhouse Damage and Benefits of Control" valued projected deaths of persons in OECD nations at $1.5 million each while deaths of persons from China, India, and Africa were valued at $150,000 each. From an ecological or human rights standpoint the discrepancy raises ethical concerns. See John Adams, "Cost-Benefit Analysis: The Problem, Not the Solution," *The Ecologist*, 26 (January/February 1996), 3.

[66] Peter G. Brown, "Toward an Economics of Stewardship: the Case of Climate," *Ecological Economics* 26 (1998), 11-21.

[67] Robert Constanza, et al., The Value of the World's Ecosystem Services and Natural Capital," *Ecological Economics* 25 (1998), 3-15 [originally published in *Nature*, 387 (May 15, 1997), 253-260]; the issue contains a number of comments on the article as well.

environmental problem lead to more uncertainty, not less. In other cases, increased knowledge about a problem leads to widening the issue, not narrowing it.

In the case of global climate change, at least three parameters help determine how one is willing to balance the knowledge-uncertainty aspect of science. These three parameters involve one's perception of the potential risk of the problem, the potential effectiveness of any reduction program, and the potential cost of the solution. If one perceives the potential risk of the problem to be slight, the potential effectiveness of any response to be questionable, and/or the potential cost to be high, one will tend to require a high threshold with respect to scientific certainty before one is willing to act. Conversely, if one perceives the potential risk to be high, the potential effectiveness of any response to be reasonable, and the potential cost to be low, one will likely be willing to act at a substantially lower threshold with respect to scientific certainty.

Each of the three lenses contributes to differing views on these parameters and on different courses of action. For example, being optimistic that energy efficiency can be gained at low cost, the technology lens can accept a somewhat lower threshold with respect to scientific certainty because the risk of high cost is discounted. Likewise, the ecological lens' concern about unintended consequences and the protection of future generations lens itself to accepting a lower threshold with respect to scientific certainty because of the precautionary need to protect the biosphere regardless of cost. In contrast, the economic lens leads one toward a cost aversion response, because the uncertainty may mean fewer benefits, a less effective response, and potentially high cost. Those viewing the issue through this lens seek more certainty before any significant investment is made in any solution.

In a study of the effects of personal beliefs and scientific uncertainty on climate change policy,[68] two researchers, Lave and Dowlatabadi, concluded that uncertainty and the degree of optimism of the decision maker were both important, but less so than whether the policymaker's decision criterion hinged on minimizing expected costs or on being as precautionary as possible. The former, criterion, focused on costs, essentially reflects the economic lens; the latter, focused on the "precautionary principle," essentially derives from the ecological lens. In a mix of scenarios, Lave and Dowlatabadi found that those focused on minimizing expected costs would most often support moderate abatement given existing uncertainties, while those focused on being precautionary would more often support stringent abatement despite costs.

This interplay of uncertainty, information, and costs is summarized in table 2. The perspective on uncertainty can have tangible policy implications – as evidenced by the ongoing debate between those who believe action to address global climate change is justified and those who do not.

---

[68] Lester B. Lave and Hadi Dowlatabadi, "Climate Change: The Effects of Personal Beliefs and Scientific Uncertainty," *Environmental Science and Technology*, Vol. 27, no. 10 (1993), pp. 1968, 1972.

Table 2. Influence of the Lenses on Policy Parameters

Approach	Seriousness of Problem	Risk in Developing Mitigation Program	Costs
Technological	By itself, the lens is agnostic on the problem. The focus of the lens is on developing new technology that can be justified from multiple criteria, including economic, environmental and social perspectives.	Believes any reduction program should be designed to maximize opportunities for new technology. Risk lies in not developing technology by the appropriate time. Focus on research, development, and demonstration; and on removing barriers to commercialization of new technology.	Viewed from the bottom-up. Tends to see significant energy inefficiencies in the current economic system that currently (or projected) available technologies can eliminate at little or no overall cost to the overall economy.
Economic	Understands issue in terms of quantifiable cost-benefit analysis. Generally assumes the status quo is the baseline from which costs and benefits are measured. Unquantifiable uncertainty tends to be ignored.	Believes that economic costs should be examined against economic benefits in determining any specific reduction program. Risk lies in imposing costs in excess of benefits. Any chosen reduction goal should be implemented through economic measures such as tradable permits or emission taxes.	Viewed from the top-down. Tends to see a gradual improvement in energy efficiency in the economy, but significant costs (quantified in terms of GDP loss) resulting from global climate change control programs. Typical loss estimates range from one to two percent of GDP.
Ecological	Understands issues in terms of its potential threat to basic values, including ecological viability and the well being of future generations. Such values reflect ecological and ethical considerations; adherents see attempts to convert them into commodities to be bought and sold as trivializing the issue.	Rather than economic costs and benefits or technological opportunity, effective protection of the planet's ecosystems should be the primary criterion in determining the specifics of any reduction program. Focus of program should be on altering values and broadening consumer choices.	Views costs from an ethical perspective in terms of the ecological values that global climate change threatens. Believes that values such as intergenerational equity should not be considered commodities to be bought and sold. Costs are defined broadly to include aesthetic and environmental values that economic analysis cannot readily quantify and monetize.

Federal Policy as Viewed through the Lenses

Faced with a fundamental problem, such as the potential for global climate change, a policymaker who is looking through the technological lens and focusing on technical fixes tends to take an activist view the government's role – to support innovation and commercialization. In the same situation, a policymaker who is looking through the economic lens and focusing on the costs and benefits of action tends to view the government's role as limited – to ensuring that any malfunctioning of the market is corrected. And a policymaker who is looking through the ecological lens and focusing on the need for action to solve the problem tends to see the government actively playing crucial roles – to inform public understanding, to seek public commitment, and to make available options for solving the problem.

These differing propensities on the role of government among the three perspectives are summarized in table 3. As described in this report, these differences have consequences for one's expectations for government action, depending on the lens one views global climate change through. At the same time, these differing expectations can have consequences for how one views the lenses themselves: that is, persons with a predisposition for limited government are likely to find the economic lens a more appropriate way to approach the issue than the other two lenses, whereas persons with a predisposition for activist government may be more comfortable with the technology and/or ecological lenses.

CONCLUSION: BALANCING THE THREE LENSES TO DEVELOP POLICY

The technological, economic, and ecological "lenses" represent ways of viewing responses to environmental problems. None is inherently more "right" or "correct" than another; rather, they overlap and to varying degrees complement and conflict with each other. Most people hold to each of the lenses to varying degrees and combinations. For example, a person who is quite concerned about the potential of global climate change form an ecological perspective, but concerned also about the economic costs and the effectiveness of a reduction program, might see a "no regrets" policy as most prudent under the circumstances. In contrast, an ecological perspective combined with a strong technological perspective would see no reason for not pushing forward with a strong reduction program without delay. A third possibility could be a risk aversion perspective deriving from cost-benefit concerns combined with a technological perspective, a combination that could lead one to a strong research and development program combined with phased-in and selective technological incentives based on potential cost-effectiveness. The combination of possibilities are many, depending on the depth of commitment to any one perspective or to any particular aspect (seriousness, effectiveness, costs) of the problem.

Table 3. Summary of Lenses

Approach	View of the Problem	Guiding Principles	Role of Government
Technological	Problem seen as opportunity for new, more efficient technology. Country seen as on the edge of an energy transition.	Technology can solve many of the problems involved if so directed.	Create market through technological mandates.
		Governmental sponsorship of and intervention in technological development can accelerate the commercialization of appropriate technology.	Economic assistance through research and development sponsored by the Government.
Economic	Problem seen in terms of internalizing a currently external cost.	The marketplace is the most efficient means of controlling undesirable pollutants.	Provide a market-based signal to private industry about the external cost (e.g., emission taxes, tradable permits, etc.)
		Private sector can solve problem given appropriate incentives with minimal governmental interference; prices are the best signal.	
Ecological	Problem seen in terms of individual and institutional behavior influenced by societal values and education.	If people have all the relevant information about choices and have the choice, they will make the responsible choice. Prices cannot signal all essential values.	Encourage a climate in which environmentally responsible decisions are more socially acceptable and less responsible decisions are stigmatized through public education and policies.
		People do not currently fully understand the implications of their behavior. The economic system and current technologies also restrict the available choices.	Ensure availability of "green" options for consumers.

Table 4. Review of Lenses Across Different Policymaking Criteria

Approach	Economic Efficiency	Effectiveness	Implementation
Technological	Depends on the cost-effectiveness of the technologies developed. Subject to considerable uncertainty during the research and development stage.	Tends to be very effective at eliminating emissions. However, the effectiveness sometimes comes at the expense of economic efficiency.	Implementation is straightforward once technology has been developed.
Economic	Depends on the functioning of the marketplace and how any economic distortions are handled.	Effectiveness depends on the level of tax/number of permits allowed and the existence of any non-market barrier to compliance	Implementation is straightforward from a governmental perspective, providing the private sector with the maximum flexibility to respond to the market's signals.
Ecological	Depends on altered values and broadened consumer choices – economic efficiency is redefined to include ecological values (such as future generations).	Can be very effective over the long-term. However, the time frame involved is unclear.	Implementation involves a combination of public education and public policy to provide consumers with the opportunity to act responsibly.

Table 3 summarizes the three lenses identified in this report. As indicated, they reflect differing assumptions about the nature of the problem, the means to a solution, and the governmental role in crafting that solution. The lenses are not mutually exclusive, but rather reflect differing emphases on what is a very complex issue.

These different emphases can be seen when examining the lenses according to different policymaking criteria; the governmental role differs substantially between the lenses. In actual implementation, any global climate change response would involve the government in multiple roles: promoting new technology, ensuring that the marketplace functions properly, and educating the public.

Table 4 presents other policymaking criteria. Once again, one sees conflict and complementarily across the different lenses. Eliminating non-market barriers can be a key to technological development, a removal that those peering through the economic lens would likely see as appropriate, although difficult. Similarly, those employing the technological lens have no objection to the ecological orientation of those using that lens, although they might question the need for such considerations – especially since those looking though the ecological lens might demand such thorough analysis of the implications of new technologies that its costs of development could be greatly increased or its adoption might be delayed. However, those viewing through the economic lens might object to the perspective given by the ecological lens, if it were to give weight to values or concerns that could not be justified through cost-benefit analysis (analysis to which those peering through the ecological lens might object).

Elements of all three lenses can be seen in the policies promoted by the Clinton Administration and in the actions of the Congress – although different perspective dominate. For the Administration, the technological (and to a lesser degree, the ecological) lens appears to dominate. The focus of Administration initiatives is on development and use of conservation and other technologies to achieve the necessary reductions without significant economic pain. That it currently does not include a massive, mandatory programs suggests that the economic lens is sufficiently powerful to prevent a strictly ecological lens from dominating the design of a climate change program. The Administration's economic analysis suggests that a flexible marketplace approach could achieve the Kyoto reduction requirements at essentially no GDP loss. The Administration does not consider costs to be the obstacle to reducing greenhouse gases that others consider it to be.

For the Congress, attention is focused on increasing certainty about the problem and the costs of actions, consistent with the economic lens. While Congress did ratify the 1992 Framework Convention on Climate Change and enacted several global climate change provisions in the 1992 Energy Policy Act, its recent actions have accentuated uncertainties and signaled a "go slow" approach. This is reflected in several actions: Both Houses have proposed to reduce funding for the Administration's technology-based Climate Change Initiative primarily on the grounds that such actions or expenses on global climate change are not yet warranted because of uncertainties concerning global warming, costs, and the commitment of other nations. During the appropriations process, the House Appropriations Committee proposed language that would have prohibited any expenditures for educating the public on global climate change (a proviso that was later struck by an amendment adopted on the floor).[69] As noted earlier, the Senate, on July 25, 1997, prior to Kyoto, agreed by a unanimous vote of 95-0 to S. Res. 98, which states the Administration should sign no agreement that would result in serious harm to the economy or that does not include developing countries (along with developed countries) within its control regime. In addition, the resolution states that any agreement submitted to the Senate include a detailed and comprehensive economic impact assessment of the treaty. The Congress' current actions on global climate change appear to focus on the issue from an economic perspective, while highlights risks of high costs, while the Administration's current activities focus on the issue mostly from a technological perspective, which discounts the risk of high costs.

Ultimately, it is the balances between all three perspectives that will shape policy options and eventually determine the character and timing of any policy response to the problem. Evolving Administration policy appears to involve incorporating an economic perspective based on tradable permits, and an ecological perspective based on increased public education. In the longer term, the Administration is suggesting that stronger measures will be necessary, reflecting an underlying ecological perspective on the issue. However, the Administration has stressed that it prefers a market-based tradable permit program to implement necessary reductions – a clear recognition of the value of the economic lens. Meanwhile, recent congressional actions have suggested that the risk of economic disruption is so high compared to the risk of global climate change – given scientific uncertainties about warming, plus uncertainties about the costs and benefits of actions to reduce greenhouse gases – that further policy development and implementation are not yet justified. (As indicated by S. Res. 98, Congress is particularly concerned about the effectiveness of any actions the United States might take if large developing nations such as China and India do not commit to specific control requirements.) Many in Congress are concerned that current efforts at technology development and public

---

[69] "The [House Appropriations] Committee is concerned that the [Environmental Protection] Agency…may be engaging in activity that is tantamount to lobbying in an effort to build public support for implementation of the [Kyoto] Protocol. While the Committee recognizes the importance of educating the public on environmental issues, there can be a very fine line between education and advocacy of an issue. The Agency…[is] thus directed to refrain from conducting educational outreach or informational seminars on policies underlying the Kyoto Protocol until…[it] is ratified by the Senate." House. Committee on Appropriations. *Departments of Veterans Affairs and Housing and Urban Development, and Independent Agencies Appropriations Bill, 1999*. House Report 105-610, p. 59.
See Dennis W. Snook, Coordinator, *Appropriations for FY1999: VA, HUD, and Independent Agencies*, CRS Report 98-204. [http://www.congress.gov/rpts/html/98-204.html#_1_4].

education may be a "backdoor" to possible implementation of a treaty that is neither justified nor ratified.

The effort by various interests to convince the public that their perspective is correct, and that those of others reflect either wishful thinking, misinformation, or excuses, will likely continue. Such efforts will be affected by improvements in the scientific understanding of global climate change, and of the domestic and international implications for strategies for addressing it. However, the pivotal decision-making point – whether that understanding warrants action or not – will be mediated in large part by the lens through which policymakers view the new knowledge.

Chapter 10

GLOBAL CLIMATE CHANGE: CARBON EMISSION AND END-USE ENERGY DEMAND

Richard Rowberg

INTRODUCTION

Background

The potential for global climate change from the buildup of greenhouse gases in the Earth's atmosphere has elicited considerable concern by the world's nations.[1] A major source of that concern is the contribution to that buildup by greenhouse gases resulting from human activity. Historically the predominant greenhouse gas from human activity has been carbon dioxide (CO_2) although other gases have been gaining in importance in recent years. Nevertheless, CO_2 is expected to remain the largest single contributor to greenhouse gas buildup from human activity for at least the next 60 years.

The principle source of such CO_2 is energy use. It is a byproduct of combustion of fossil fuels. In 1996, the Department of Energy (DOE) estimated that the United States produced 1,462 million metric tons-carbon equivalent (MMTCE)[2] of CO2 from the combustion of 85.2 quadrillion BTUs (Quads) of fossil fuels.[3]

The growing concern about greenhouse gas-induced global climate change has prompted major international efforts to limit the buildup. The most recent of these was the December 1997 United Nations (Kyoto) Protocol on Global Climate Change, which established greenhouse gas reduction targets for the developed nations signatory to the

---

[1] Congressional Research Service, *Global Climate Change,* by Wayne Morrissey and John Justus, CRS Issue Brief 89005 (updated regularly).

[2] One ton-carbon equivalent is equal to 3.67 tons of carbon dioxide.

[3] Energy Information Administration, Department of Energy, *Annual Energy Outlook, 1998: With Projections Through 2020*, DOE/EIA-0383(98) (December 1997), 100, 124.

accord.[4] That agreement, which the United States has signed but not ratified, would require the United States to achieve average annual carbon-equivalent emissions of six greenhouse gases over the period 2008-2012 that are 7% below specified baseline years. For CO_2, the baseline year is 1990. According to DOE, the United States produced 1,353 MMTCE in 1990 from energy use.

There have been several analyses of the implications of meeting the reduction targets set by the Kyoto accord. Most recently, the Energy Information Administration (EIA) of DOE reported on the impacts on the U.S. energy markets and economy of those reductions.[5] In addition, a DOE study carried out by five of its national laboratories examined the potential for new energy supply and demand technologies to help meet those targets with a minimum of economic dislocation.[6] Those studies carried out detailed examinations of U.S. energy demand in analyzing the potential impacts of CO_2 emission reduction.

With some exceptions, however, the detail did not extend to the specific end-uses – such as motor vehicles or air conditioning – that consumers are familiar with. Although the Inter-Laboratory (called the 5-lab) study looked at specific end-use technologies, it did not provide data on end-use energy demand and CO_2 emissions for all major end-uses. While such information is not necessary to estimate potential economic impacts of CO_2 emission reduction, it can be helpful in gaining an understanding of the potential consequences of such actions at the consumer level.

Report Purpose and Format

This report presents estimates of actual and forecast energy demand for all of the common energy demand end-use categories for 1996, 2008, and 2012. It then calculated the CO_2 emissions for these estimates. With these values the reader should be able to see which, of the several ways energy is used in this country, are the major sources of CO_2 emissions. A similar report was prepared in 1991, which presented energy demand and CO_2 emissions estimated for 1988 and 2000.[7]

This report goes further than the previous report by presenting an analysis of the CO_2 emission reductions called for by the Kyoto accord. This analysis is based on a spreadsheet model that estimates the CO_2 emissions reduction for each of the end-use categories for both 2008 and 2012, and calculates the energy demand reductions needed to meet those emissions targets. In this way, the reader can see just how any of the common energy demand end-uses would be affected by reaching the targets. The report

---

[4] Congressional Research Service, *Global Climate Change Treaty: Summary of the Kyoto Protocol*, by Susan Fletcher, CRS Report 98-2, 22 December 1997.

[5] Energy Information Administration, Department of Energy, *Impacts of the Kyoto Protocol on U.S. Energy Markets and Economic Activity*, SR/OIAF/98-03, (October 1993).

[6] Office of Energy Efficiency and Renewable Energy, Department of Energy, *Scenarios of U.S. Carbon Reductions: Potential Impacts of Energy Technologies by 2010 and Beyond*, prepared by the Inter-laboratory Working Group on Energy-Efficient and Low-Carbon Technologies.

[7] Congressional Research Service, *Energy Demand and Carbon Dioxide Production*, by Richard Rowberg, CRS Report 90-204, 11 February 1991.

concludes with a discussion of implications of these reductions for representative end-uses.

ENERGY DEMAND BY END-USE CATEGORY

Background

The EIA divides U.S. energy demand into four sectors: industrial, residential, commercial, and transportation. Each sector is characterized by sources of energy supply and by specific end-uses. The sources of energy are the fossil fuels, renewables, and electricity. Electricity, of course, requires primary energy sources, namely fossil fuels, nuclear fuels, and renewables. End-uses are defined as specific functions or equipment that perform services for consumers such as cooking, heating, rail transport, electric motors, and lighting. Table 1 presents the energy sources and end-uses considered in this report. There are 27 end-used in all. The term in parentheses after each end-use is a symbol used to denote the end-use in graphical representations to follow. The end-uses are not presented in any special order. Also, the energy sources in the right column are not meant to correspond to the end-uses in the left column. A given end-use usually requires more than one source of energy.

While most of the end-uses are self-explanatory, some need further explanation. The miscellaneous end-use in the residential sector includes motors and heating elements commonly found on gardening equipment, machine tools, and small appliances. Appliances in this sector include refrigeration, cooking, freezers, clothes washers and dryers, color televisions, and personnel computers. In the commercial sector, miscellaneous includes electronic office equipment, telecommunications, medical equipment, service station equipment, and manufacturing performed in commercial buildings.

In the industrial sector, direct heat refers to manufacturing processes requiring the direct application of heat such as metal treating and chemical production. Machine drive refers primarily to the use of electric motors to control manufacturing processes. Steam is used in manufacturing primarily as a heat source for chemical and metallurgical processes, as a power source for metal-forming equipment, and to drive turbines. Electrolysis is an electro-chemical process used primarily in aluminum production.

In addition to those manufacturing processes, the industrial sector contains construction, agriculture, and mining. Each of these "sub-sectors" contains several end-uses, but lack of good data prevents an accurate disaggregation. Therefore, each is counted as a separate end-use. Finally, the industrial sector uses a significant quantity of petroleum fuels for products such as asphalt, liquefied petroleum gases, petrochemical feedstocks, and lubricants. Most of the carbon in these non-energy applications is not released as CO_2 emissions, and the non-energy end-uses are not included in the carbon emission analysis to follow.[8]

Table 1. End-Uses and Energy Sources (By Sector)

End-uses	Energy Sources
Residential	
Space Heating (SHR) Water Heating (WHR) Appliance (ApR) Air Conditioning (ACR) Lighting (LR) Miscellaneous (MR)	Distillate Fuel Oil Liquid Petroleum Gas Natural Gas Coal Renewable Energy Electricity
Commercial	
Miscellaneous (MC) Lighting (LC) Air Conditioning (ACC) Space Heating (SHC) Water Heating (WHC) Cooking and Refrig (CRC)	Distillate Fuel Oil Residual Oil Liquid Petroleum Gas Motor Gasoline Natural Gas Renewable Energy Electricity
Industrial	
Direct Heat (DHI) Machine Drive (MDI) Steam (StI) Construction&Agri (CAI) Mining (MnI) Heat & Air Cond (HAI) Electrolysis (EI) Lighting (LI) Electric Gen (EGI)	Distillate Fuel Oil Liquid Petroleum Gas Residual Oil Motor Gasoline Natural Gas Coal Renewable Energy Electricity
Transportation	
Light Duty Vehicles (LDT) Freight Trucks (FTT) Air (ArT) Marine (MaT) Rail (RT) Pipeline (PT)	Distillate Fuel Oil Jet Fuel Motor Gasoline Residual Oil Liquid Petroleum Gas Natural Gas Renewable Energy Electricity

---

[8] A small fraction of the carbon - about 20% - of these non-energy uses, according to the EIA, does end up as CO_2 in the atmosphere. The effect of this contribution will be discussed but will not be included in the detailed results because it is very small. See: Energy Information Administration, Department of Energy, *Emissions of Greenhouse Gases in the United States, 1996*, DOE/EIA-0573(96), (October 1997), 70.

In the transportation sector, light duty vehicles include all vehicles weighing less than 8,500 pounds. These include automobiles, minivans, sport utility vehicles, larger passenger vans, and small trucks. Freight trucks include all commercial and freight trucks weighing over 8,500 pounds

In each sector, electricity is given as one of the energy sources. Electricity, of course, requires primary energy sources for its generation. These sources are distillate and residual fuel oil, natural gas, coal, and nuclear and renewable energy. In the calculations to follow, it is assumed that for a given year, the mix of energy sources to generate electricity is the same regardless of which sector uses the electricity. The mix will change over time, however, and those changes are incorporated.

Analytical Method

Energy demand in the United States for 1996 was adopted for the baseline because that is the last year for which complete data are available for all sectors. For the residential and commercial sectors, data for end-use energy demand by fuel are directly available from the EIA.[9] Consolidation of some of the end-uses provided by EIA was made by combining several household appliances in a category called residential appliances, and combining all commercial categories labeled other uses with commercial office equipment in a category called commercial miscellaneous. These new categories are described above.

For the transportation sector, the EIA provides separate reporting of total energy use by the various categories and by energy source.[10] As a result, estimates have to be made of how much of a given energy source are used by any end-use. This can be done by noting that there is a predominate energy source for a given end-use and that most of the categories will use no more than two different kinds of energy sources. For example, light duty vehicles will use primarily gasoline with a small amount of distillate, air transport will use all of the jet fuel and a small amount of gasoline, rail is the primary user of electricity but also uses residual fuel oil. By reconciling the total energy demand for each end-use with that for each energy source, an accurate picture of energy demand by energy source for each end-use can be obtained.

For the industrial sector, the calculation is more complex. The sector is made up of four subsectors: manufacturing, construction, agriculture, and mining. Detailed energy use data by end-use exist only for the manufacturing sector. The EIA publishes a survey of manufacturing energy use every three years, the most recent for 1994. The report provides data on the major end-uses by energy source.[11] A complication arises in that a large fraction of the totals reported by end-use and by energy source are not specified. By going to the individual industry groups much of those unspecified values can be estimated from noting the types of processes used by those industries. For example,

[9] Energy Information Administration, *Annual Energy Outlook 1998*, 106-109.
[10] Ibid. 102, 111.
[11] Energy Information Administration, Department of Energy, *Manufacturing Consumption of Energy, 1994*, DOE/EIA-0512(94), (December 1997), 114.

unspecified fuel in the paper and paper products industry is likely to be wood used to produce steam. In petroleum refining, the unspecified fuel is likely to be still gas, a product of the refining process, used for direct heat and steam. Once energy demand by end-use and energy source is determined for 1994, the 1996 value can be estimated by adjusting each energy source by the 1994-1996 growth rate. That method assumes that there is no significant shift of the type of energy source used by the end-uses over that period.

For mining, the census of mineral industries published by the U.S. Census Bureau of the Department of Commerce provides data on energy use for1992 by energy source for the entire mining industry.[12] No data are provided by end-use. For that reason, the entire mining industry is included as a separate end-use. To obtain values for 1996, the value of each energy source is adjusted by its 1992-1996 growth rate. For construction and agriculture, no energy use data are available. Total energy demand by energy source for the entire industrial sector, however, is available.[13] By subtracting total energy demand by energy source for the other two sub-sectors from the industry total, and considering construction and agriculture as one end-use, its energy demand can be estimated as the residual.

For the years 2008 and 2012, the EIA forecasts are used to estimate energy demand by end-use and energy source.[14] *The Annual Energy Outlook, 1998* provides forecasts for the years 2005, 2010, and 2015, among others.[15] To obtain forecasts for 2008 and 2012, therefore, it is first necessary to determine the forecast for energy demand by end-use for 2010.[16] Energy demand forecasts for the residential and commercial sector end-uses are obtained directly for 2010 just as described above for 1996. Similarly, estimated energy demand by end-use for the transportation sector for 2010 can be obtained from the EIA data using the same methods as for 1996.

For industry, a different method must be used because there are no forecasts of manufacturing energy demand by end-use. The EIA, however, does provide forecasts of energy demand by energy source for the entire industrial sector. By adjusting each energy source for a given end-use by the 1996-2010 growth rate for that energy source for the entire sector, an estimate of the energy demand for that end-use for 2010 can be made. This method assumes that the relative mix of energy sources for a given end-use does not change significantly over that period. That is a valid assumption because a major energy-source mix change in one end-use would require an equal and opposite change in one or more of the remaining end-uses in order to keep the totals for the sector unchanged. Such actions are quire unlikely, particularly over a 14-year period.

[12] U.S. Census Bureau, Department of Commerce, *1992 Census of Mineral Industries: Fuels and Electric Energy Consumed*, MIC92-S-2. See: [http://www.census.gov/mcd/minecen/download/nc92feec.text].

[13] Energy Information Administration, *Annual Energy Outlook, 1998*, 101.

[14] For this report, the EIA reference case forecast is used. That forecast is based on a macroeconomic model that calculates a series of indicators that are used to drive the energy demand model. Among the indicators for the reference case are an annual growth of real GNP of 3.0% per year between 1996 and 2020, and a decline in total energy intensity (1000 BTU/1992 dollar of GDP) of 0.9% per year over the same period. See, Energy Information Administration, *Annual Energy Outlook, 1998*, 125.

[15] Energy Information Administration, *Annual Energy Outlook, 1998*, 101, 106-111.

[16] In all cases the EIA reference case forecast is used.

Once the forecasts for 2010 are obtained, they can be modified to provide estimates of the forecasts for 2008 and 2012. This modification is performed by first calculating the annual growth rate between 2005-2015 for each energy source by sector. The data for these calculations are obtained from the EIA *Annual Energy Outlook, 1998*. With those growth rates, it is a straightforward matter to adjust each of the energy sources for each of the end-uses by the appropriate annual growth rate. This method assumes that the 2005-2010 and 2010-2015 growth rates for a given energy source are the same for all end-uses in a sector. In the cases where that assumption can be checked – the residential and commercial sectors – it is not strictly correct. The error introduced for those two sectors, however, will be small because the growth rates themselves are small, 1% per year and less. For the other two sectors, it is unavoidable, given the data available, but also quite small.

Results

The results of those calculations are shown in Figure 1 (on next page) for all 27 end-uses. They are arranged in descending order of energy demand forecast in 2012. The results present primary energy demand for each end-use; waste heat produced in the generation of electricity is assigned to the electricity total for each end-use.[17] Tables presenting the actual data are in the Appendix to this report.

Figure 1. Energy Demand by End-use: 1996, 2008, 2012

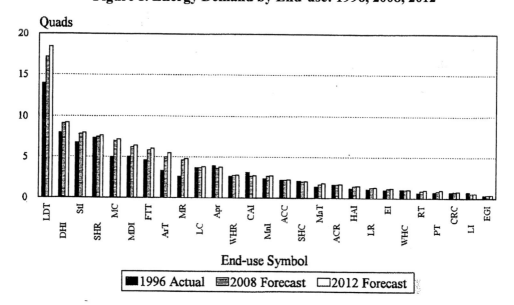

---

[17] About two-thirds of the energy used to produce electricity is lost as waste heat. Thus one Quad of electric energy delivered to consumers represents about three Quads of primary energy used. Some of the waste heat is used for low-level heat – e.g., space heat – in industry and commercial buildings, and, as such, replaces other energy sources. This secondary use of "waste heat" is particularly prevalent with electricity generated on-site by industry. This process is called cogeneration.

Total energy demand represented by all 27 end-uses in 1996 – 87.5 Quads – is nearly equal to the value of total energy demand reported by EIA – 88.1 Quads – indicating that these 27 end-uses virtually capture the full range of U.S. energy demand. For 1996, EIA also reports that non-energy uses, as discussed above, consumed 5.95 Quads of fossil fuels. For 2008, energy demand for the 27 end-uses is 101.5 Quads and for 2012 it is 105.7 Quads. The EIA estimates for total energy demand including non-energy uses of fossil fuels for those two years are 109.6 Quads and 113.6 Quads, respectively. These values would indicate that non-energy use for 2008 and 2012 is 8.1 Quads and 7.9 Quads respectively. Based on extrapolation of past trends those values are reasonable, although it is unlikely that non-energy, fossil fuel use would decline from 2008 to 2012. The value for 2008 is probably a little high, while that for 2012 somewhat low. This means that the estimates for energy demand determined from summing the end-uses is somewhat low for 2008 and somewhat high for 2012. The discrepancy is small, however, and very likely to be within the EIA forecast errors.

Of the individual end-uses, the one with the largest energy demand is light duty vehicles. It is nearly twice the size of the next largest end-use, direct hear for industrial processes. Furthermore, its energy demand is forecasted to grow significantly from 1996 to 2012 primarily as a result of a substantial increase in vehicle miles traveled.[18] Other end-uses that are expected to grow sharply between 1996 and 2012 are the miscellaneous categories in both the residential and commercial sectors. A rapid expansion in office electronic and telecommunication equipment, and in a variety of small residential appliances and outdoor heating equipment and motors is expected to be responsible for that growth. Electricity is by far the dominant energy source for those miscellaneous categories – constitute 48% of the projected growth in energy demand between 1996 and 2012.

CO$_2$ EMISSIONS

Analytical Method

The first step in estimating CO$_2$ emissions is to determine the carbon emission coefficients for the fossil fuels used as energy sources. Each fossil fuel has a characteristic CO$_2$ production rate, or emission coefficient, determined by the chemistry of that fuel. That rate is the amount of CO$_2$ that will be produced upon complete combustion of a specific quantity of fuel. Those rates are given in Table 2 in term so of millions of metric tons of carbon (MMTC) produced per Quad of energy used.[19],[20] The coefficients for a given fuel can change from year-to-year depending on the quality of the fuel produced that year. The changes will be small, however, and the EIA has reported no

[18] Energy Information Administration, *Annual Energy Outlook, 1998*, 111. EIA forecasts only a small increase in light duty vehicle energy efficiency over that period: 20.2 mpg in 1996 to 20.3 mpg in 2010.
[19] Energy Information Administration, *Emissions of Greenhouse Gases*, 100.

changes for petroleum products and natural gas over the period 1986 to 1996.[21] There have been slight changes for coal, but they are very small and possible future changes will not be considered in this report. Therefore, the carbon emission coefficients for fossil fuels are assumed to remain constant over the period covered by this report, 1996 to 2012.

Table 2. Carbon Emission Coefficients (Millions of Metric Tons per Quad)

Source	1996	2010
Gasoline	19.38	19.38
LPG	16.99	16.99
Jet Fuel	19.33	19.33
Distillate	19.95	19.95
Residual	21.49	21.49
Coal – Res/Com	26.00	26.00
Coal – IND	25.63	25.63
Natural Gas	14.47	14.47
Electricity	16.04	16.50

For electricity, the coefficient is calculated from the coefficients of the fossil fuels used in the generation mix, weighted by their contribution to that mix. This mix will change over time. Using EIA forecasts for the generation mix in 2010, the carbon emission coefficient for electricity for that year can be calculated. Note that the change between 1996 and 2010 is small. While the value for 2008 and 2012 can be estimated by using the 2005-2010 and 2010-2015 fossil fuel growth rates given by the EIA, the changes from 2010 are likely to be insignificant. Therefore, the 2010 value is used in calculating carbon emissions for 2008 and 2012.

Once the carbon emission coefficients are calculated, it is relatively straightforward to calculate the total amount of carbon emitted by each end-use in the selected year. For each end-use, the quantity of energy from each energy source is multiplied by the appropriate carbon emission coefficient and the results are summed. Using those coefficients, of course, will yield carbon emissions not CO_2 emissions. Although it is simple to convert to CO_2 emissions (see footnote 19), this will not be done in order to be consistent with the way the EIA presents emission data.

Results

The results of this calculation are given in Figure 2.

---

[20] These coefficients are presented in terms of the amount of carbon produced. To calculate the amount of CO_2 produced it is necessary to multiply the coefficient by 3.67, the ratio of the molecular weight of CO_2 to that of carbon.

[21] Energy Information Administration, *Emissions of Greenhouse Gases*, 100.

Figure 2. Carbon Emissions by End-use: 1996, 2008, 2012

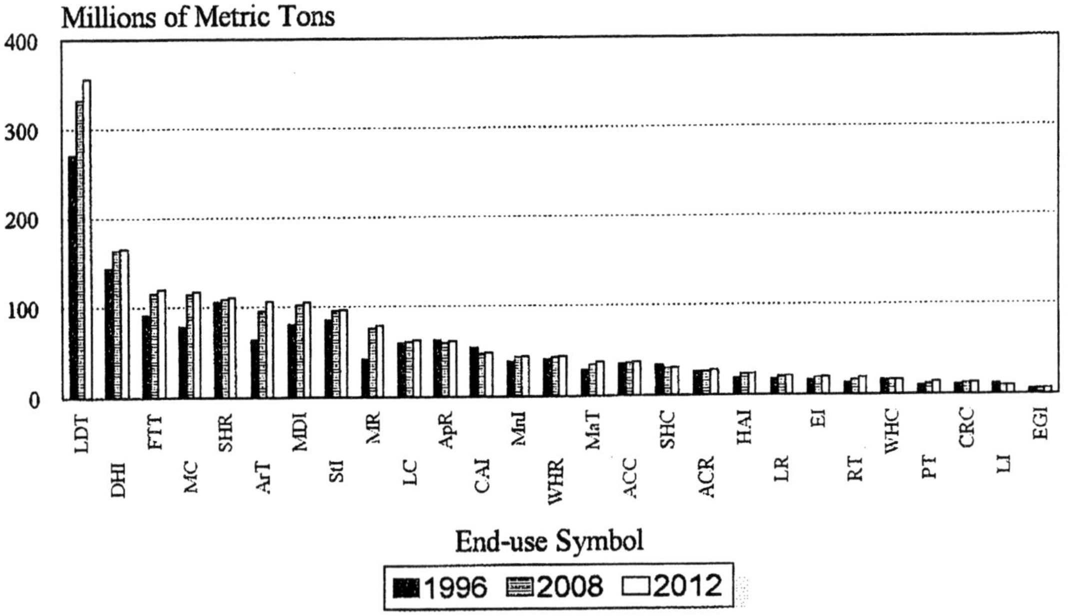

The detailed data are provided in tables in the Appendix. The data in the figure are displayed by order of end-use emission rankings forecast for 2012. Total carbon emissions in 1996, calculated by aggregating end-uses, were 1,464 million metric tons carbon equivalent (MMTCE). The EIA reports that total 1996 carbon emission, including those from non-energy use of fossil fuel, were 1,463 MMTCE.[22] The CRS estimate might be slightly high because no emissions from non-energy uses of fossil fuels were included. As noted above, about 20% of the carbon in those fuels was released to the atmosphere. This would amount to carbon emissions of about 15 to 20 MMTCE. Nevertheless, the value calculated here is quite close to the EIA total indicating that the end-use structure is valid.

For 2008 and 2012, aggregation of the end-use carbon emissions yields totals of 1,721 and 1,795 MMTCE respectively. The EIA forecasts for these two years are 1,757 and 1,837 respectively. Considering the effect of non-energy uses of fossil fuels as discussed in the previous paragraph, the results obtained from aggregating end-use emissions appear quite consistent with the EIA forecasts.

The largest contributor to carbon emissions was, and is expected to continue to be, light duty vehicles. That end-use if forecast to contribute almost twice as much carbon as the next highest end-use, direct heat for industry. These observations, of course, are consistent with those reported above in the discussion of energy demand. There is not,

---

[22] Energy Information Administration, *Annual Energy Outlook, 1998*, 124.

however, a direct correlation between total energy demand and carbon emissions. Rather the energy source mix can play an important role in determining the relationship between energy demand ranking and carbon emission ranking. In particular, industrial steam and residential space heat rank higher on the energy demand scale than the carbon emission scale because both include a significant amount of biomass – wood combustion – in their energy source mix. Wood combustion does not produce any net production of CO_2 as long as the wood used for energy production was not previously counted as a sink for CO_2.[23]

Nevertheless, there are a number of similarities in the two tables. Light duty vehicles, and residential and commercial miscellaneous are projected to have the largest increases in carbon emission between 1996 and 2008, and 1996 and 2012. Other end-uses that indicate a large increase are freight trucks, air transport, and industrial machine drive. These five end-uses make up 70% of the forecast increase of total carbon emissions from energy use between 1996 and 2012.

For some end-uses, carbon emissions are forecast to change very little and, in some cases, decline. All of the end-uses in the residential and commercial sector, except the miscellaneous categories, are expected to be nearly the same in 2012 as in 1996. This level behavior is due primarily to projected increases in energy efficiencies for those end-uses. Most industry and transportation end-uses, however, are forecast for significant growth. On average, the rate of forecast efficiency gains for the end-uses in those sectors does not match their projected energy demand growth resulting from continued economic growth.

EMISSION REDUCTIONS

Kyoto Protocol Targets

In 1997, the United States joined with more than 160 nations to negotiate greenhouse gas emission reductions targets. The result was the Kyoto Protocol that established such targets for the Annex I countries.[24] Those countries are among the ones that have ratified the 1992 United Nations Framework Convention on Climate Change. According to the Protocol, the United States is to reduce those emissions to a level 7% below 1990 levels. There are six greenhouse gases covered in the Protocol and the targets are based on the carbon equivalent of each gas. These gases are CO_2, methane, nitrous oxide, hydro fluorocarbons, per fluorocarbons, and sulfur hexafluoride. For the last three gases, there is an option that 1995 can be adopted as the base year.[25]

---

[23] During tree growth, carbon is sequestered as a result of the photosynthesis process whereby plants consume CO_2 and give off oxygen. Because this growth takes place within years of the time when the combustion of the wood takes place, there is no net production of CO_2 over the time frame of concern for possible greenhouse gas induced climate change.

[24] Congressional Research Service, *Global Climate Change Treaty: Summary of the Kyoto Protocol*; and Congressional Research Service, *Global Climate Change*.

[25] Energy Information Administration, *Impacts of the Kyoto Protocol on U.S. Energy Markets and Economic Activity*, xi.

In 1990, total emissions of the six greenhouse gases was 1,618 MMTCE.[26] If the 1995 base year is adopted for the three gases, the total becomes 1,629 MMTCE. Of the total, 1,374 MMTCE is CO_2, of which 1,346 MMTCE comes from fossil fuel combustion for energy use. The non-energy CO_2 is from certain industrial processes, primarily cement production and limestone consumption.

There are considerable uncertainties involved in calculating the targets and reduction quantities required.[27] One of the principal ones is how the reduction will be allocated among the various gases. It might be possible to reduce some gases well beyond the 7% target allowing smaller reductions in others. In addition, the Protocol allows countries to trade emissions credits.[28] That process could allow a loosening of the target level for the United States.

One of the purposes of this report is to show what would happen to energy demand levels for the common end-uses if the targets were applied uniformly. Therefore, to determine the required carbon emission targets, the 7% reduction will be taken from carbon emissions resulting from energy production in 1990. This target level is 1,252 MMTCE. One other case has been considered, that of a 3% reduction from the 1990 levels. This reduction level was estimated by the U.S. Department of State and cited by the Council of Economic Advisors as possible given the flexibility inherent in the Protocol.[29] This target level is 1,306 MMTCE. The difference between the two target levels is small, however, and will not significantly affect the implications of reaching these levels for any of the end-uses.

In addition to determining the target level, there is also the question of when the reductions would start. The Protocol states that the average emissions over the five-year period must equal the target. Therefore, if the 7% reduction is met in 2008, emissions can remain flat throughout that period. If no reduction has taken place by 2008, the reduction necessary by 2012 must be substantially greater. This behavior is shown in Figure 3, which compares different trajectories to the target level. In the extreme case of waiting until 2008 to begin, and assuming a constant percentage decline each subsequent year, the 2012 target level becomes 756 MMTCE, 59% below the current EIA forecast for that year. Starting in 2005, the case adopted by the EIA assessment,[30] and assuming a constant percentage decline to the target of 1,252 MMTCE in 2008, would require a 6.9% per year rate of decline. Starting in 1999 to the same target, would require a 1.9% per year decline.

It is clear that the degree of difficulty in reaching the target levels would increase dramatically as the year in which the reductions begins approaches 2008. For the purposes of this paper, it is assumed that reduction would begin some time before 2008 and reach the 7% (or 3%) level by 2008 so that carbon emissions are constant over the 2008-2012 period.

---

[26] Energy Information Administration, *Emissions of Greenhouse Gases*, x, 18.

[27] For an extensive discussion of the uncertainties, see, Congressional Research Service, *Global Climate Change: Reducing Greenhouse Gases – How Much from What Baseline?* By Larry Parker and John Blodgett, CRS Report 98-235 ENR, 11 March 1998.

[28] Congressional Research Service, *Global Climate Change*.

[29] Energy Information Administration, *Impacts of the Kyoto Protocol on U.S. Energy Markets and Economic Activity*, xii.

[30] Ibid.

Figure 3. Carbon Emission Reduction Projections

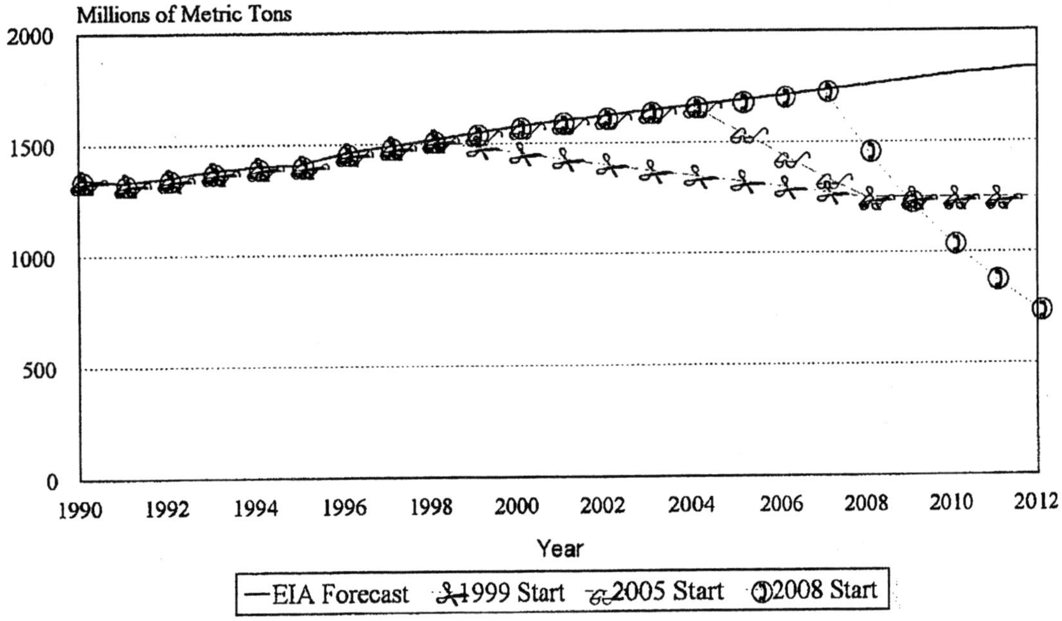

Reduction Levels

To estimate the consequences of reaching the reduction targets of 7% below 1990 carbon emissions on end-use energy demand in 2008, it is assumed that the reductions are applied uniformly to all the end-uses. The 2008 carbon emissions of each end-use are multiplied by the ratio of the target emissions level to the total carbon emissions from all energy use in 2008 as forecast by the EIA.[31] The results are the allowed carbon emissions for that end-use to meet the targets set by the Kyoto Protocol. Once those target emission levels are determined, the energy demand levels that would be required to produce those target emission levels can be calculated. First, the average carbon emissions coefficient for each end-use is calculated using the 2008 values.[32]

---

[31] The EIA forecast is used instead of the aggregate from all of the end-uses calculated by CRS because the former accounts for the non-energy contribution as described above, and such contributions are included in the target level. Therefore, both parts of the ratio will be comparable.

[32] This is just the ratio of total carbon emissions for the end-use to the total energy demand for that end-use from all energy sources.

Figure 4. End-use Energy Demand Reduction Requirements

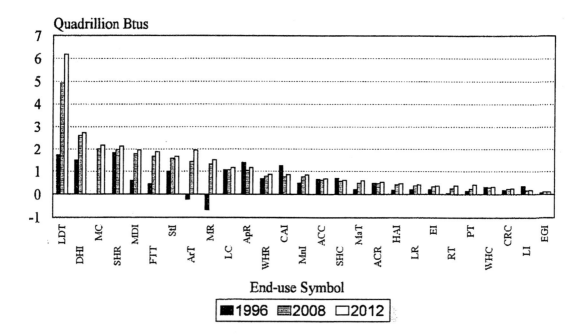

Then, the target emission levels for a given end-use are divided by its average carbon emission coefficient, giving the target energy demand level. Finally, the difference between the forecast level and the target level gives the required energy demand reduction needed to meet the Kyoto Protocol targets for each end-use.

The results of this calculation are shown in Figure 4 for 2008. The detailed data are in the Appendix. Because carbon emissions would have to remain constant at 1,252 MMTCE over that period to meet the Kyoto Protocol requirements, energy demand for each end-use would also have to remain constant as long as each end-use is sharing the reduction proportionately. The higher energy demand forecasts for 2012 as shown above mean that the reductions for each end-use in 2012 would be correspondingly higher. These reductions are also shown in Figure 4 and the detailed data are shown in the Appendix.

Figure 4 also presents the energy demand reductions from 1996 actual levels that would result from meeting the Kyoto Protocol requirements. A negative value means that the target energy demand level in the 2008-2012 period would be higher than the 1996 actual level. Only two end-uses show such behavior, residential miscellaneous and air transport. For both, EIA forecasts that energy demand is expected to grow well above the average of all end-uses. Other end-uses that are forecast to grow rapidly, such as commercial miscellaneous and light duty vehicles, show relatively small reductions from the 1996 actual levels. For nearly all of the end-uses, however, the target levels for 2008-2012 would be significantly below the 1996 recorded energy demand. Indeed, some end-uses show larger reductions from 1996 levels than from the 2008/2012 forecasts. Those are primarily in the residential and commercial sectors and are end-uses where substantial efficiency gains are expected in coming years such as residential appliances and commercial space heating. In percentage terms, the reductions that would be required from 1996 actual levels range from about a negative 25% to a positive 49% with an average of about a positive 20%.

Under the method used in this report to apportion the reductions that would be required by the Kyoto Protocol, each end-use contributes the same percentage reduction in carbon emissions and energy demand for 2008. For that year, energy demand would be reduced by about 28.7%. Because the energy demand target level remains fixed from 2008 to 2012 but demand growth for each end-use is forecast to grow at different rates over that period, the percentage reduction in energy demand for 2012 would not be the same for each end-use. The percentage reductions for 2012 in end-use energy demand would range from about 29% to 41% with most around 31%.

Discussion

Implications. Reaching the Kyoto Protocol target for carbon emissions from energy use would result in energy demand in the year 2008 of about 70.2 Quads.[33] This would be about 28.3 Quads below the amount used in 1996. It would be the lowest total U.S. energy demand since 1988. Taken from the 1996 level, to reach that target by 2008 would require a decline of about 1.6% per year. If actions do not begin until 2005, as assumed in the EIA study on impacts of the Kyoto Protocol, energy demand would have to decline by about 10.5% per year to reach the target level. The longest stretch of declining energy demand in the United States since 1949 occurred from 1979 to 1983 following the second Arab oil embargo and oil price spike. Over that period, total energy demand dropped by 10.6% or about 2.8% per year for the four-year period. While that rate is considerably greater that 1.6% per year, it is much less than 10.5% per year, and the total percentage drop that took place is about one-half that which would be required to meet the 2008 Kyoto Protocol target from 1996.

In discussing these results, three examples are considered: light duty vehicles, residential space heating, and industrial direct heating. Those are large energy demand

---

[33] In addition, there would be about 7.5 Quads of fossil fuel use for non-energy purposes and about 3.6 Quads of biomass making a total of about 81.3 Quads.

categories in three different sector. The assessment will consider changes in energy efficiency needed to meet the targets as well as consequences of reducing energy demand without any efficiency gains. Finally, a review of historical energy demand for each of the three end-uses will be made to see whether there are precedents for reductions called for to meet the Kyoto Protocol targets.

Light Duty Vehicles. The end-use with the largest energy demand is light duty vehicles. Meeting the Kyoto target would require a decline in energy use by these vehicles of 4.9 Quads from the level now forecast by the EIA for 2008 and about 1.7 Quads from the amount used in 1996. This would amount to a drop of about 880,000 barrels a day of gasoline consumption. To achieve that decline by 2008 would require an increase of the average fuel efficiency of the entire light duty vehicle fleet from 20.3 miles per gallon (mpg) to 23.3 mpg assuming *no* increase in vehicle miles traveled (VMT) between 1996 and 2008. If the EIA forecast of VMT for 2008 is met, fuel efficiency would have to reach 29.6 mpg, an annual rate of increase of 3.2%. Currently, the EIA forecasts that the light duty vehicle fleet will operate at 20.3 mpg in 2008.

While historical fuel efficiency data for light duty vehicles as a group do not exist, data for passenger cars, which constitutes about 60% of all light duty vehicles, show that fuel efficiency grew from 14.6 mpg in 1979 to 21.2 mpg in 1991. The annual rate of increase for that 12 year period was 3.2%.[34] For all motor vehicles, however, the increase was considerably less over that period, 2.6% per year. Because the latter includes freight trucks, the actual increase for light duty vehicles was somewhere in between these two rates. Since 1989, however, fuel efficiency for light duty vehicles has increased from 18.5 mpg to 20.2 mpg. This slow growth in fuel efficiency plus the forecast that such growth will continue to be slow for the next two decades at least, indicates that most of the effects of the fuel economy standards established in the late 1970s might already have been achieved. Therefore, to achieve the growth in fuel economy required to meet the Kyoto target, would require a return to rapid growth in new vehicle fuel economy.

It should be noted, however, that since the mid-1980s, there has been little increase in the Corporate Average Fuel Economy (CAFÉ) standards for new automobiles or for light trucks such as sport utility vehicles, vans, and pickup trucks. The CAFÉ standards for new automobiles has remained at 27.5 mpg since 1985, and the standard for new light trucks has only increased from 20.5 mpg in 1987 to 20.7 mpg in 1996.[35] Furthermore, over the same period the automobile portion of the light duty vehicle fleet has been decreasing. In 1979, over 79% of all vehicle miles traveled by light duty vehicles were accounted for by automobiles. In 1996, that percentage had dropped to just over 64%.[36] The remainder of those miles were accounted for by light trucks, which are less fuel-

---

[34] Energy Information Administration, Department of Energy, *Annual Energy Review 1997*, DOE/EIA-0384(97) (July 1997), 53.

[35] National Highway Transportation Safety Administration, Department of Transportation, *Automobile fuel Economy Program: Twenty-second Annual Report to Congress, Calendar Year 1997*, 3. See: [http://www.nhtsa.gov/cars/problems/studies/fuelecon/index.html].

[36] Oak Ridge National Laboratory, Department of Energy, *Transportation Energy Data Book*: Edition 18, ORNL-6941 (1998), 5-5. The report can also be found on [http://www-cta.ornl.gov/data/tedb18/Index.html].

efficient than automobiles. In 1996, the value for all automobiles was 21.5 mpg and for all light trucks was 17.3 mpg.[37]

If no efficiency gains were to take place, a substantial decline VMT would be necessary. At a constant 20.2 mpg, vehicle miles traveled would have to decline about 12.4% from the 1996 actual level and about 28.8% from the level forecast for 2008. In terms of automobiles which make up about 72% of the VMP, this reduction would force annual miles traveled per automobile to drop to about 10,300 compared to the 1996 level of about 11,700 miles.[38] If miles traveled per automobile per year increased at the same rate as VMT according to the EIA forecast, the value in 2008 is now projected to be about 14,500 miles. Therefore, the VMT level required to meet the Kyoto Protocol target without any efficiency gains would be substantially lower.

Residential Space Heating. To meet the Kyoto targets for residential space heating energy demand, a decline would be required of 1.8 Quads, or 27.6%, from the 1996 level and about 2 Quads, or 28.7% from the level now forecast by EIA for 2008. The small difference between the 1996 and 2008 reductions is a result of the continued efficiencies that the EIA forecasts will be forthcoming for residential space heating.

To achieve those reductions would require large increases in the efficiencies of residential building shell and/or heating systems, a substantial reduction in the temperature levels maintained in a typical house, or some combination of both. For example, if a typical residential building had an average building thermal barrier rated at R=16, it would have to increase to R=22 to reduce its space heating energy requirements for heating by 27.6%. It could also install new heating equipment that operated at a higher efficiency. To achieve the necessary fuel reduction, an efficiency gain of over 37% would be needed. For example, a furnace operating at 60% efficiency would need to be replaced by one operating at about 82%. Finally, if neither of these two efficiency improvements were possible, that residential building would need to decrease its average indoor temperature. For example, if the building's normal indoor temperature was 75 degrees, and the outdoor temperature averaged 40 degrees, the indoor temperature would have to be lowered to about 65 degrees to reduce its heating requirements in line with the Kyoto target.

A change in residential space heating energy demand of this magnitude has occurred before, form 1978 to 1980. the large increase in energy prices in the late 1970s drove the demand for space heating energy down by 22%, after correcting for changes in heat load, over that two-year period.[39] After that decline, however, residential space heating energy demand has stayed nearly constant, increasing by less that 10% between 1980 and 1996 after correcting for changing heating requirements. That behavior is due primarily to increasing building and heating system efficiencies, which have kept pace with housing stock growth and declining real energy prices.

DOE has assumed that those efficiency increases will continue in order to compensate for future housing stock growth. DOE forecasts that space heating demand

---

[37] Ibid., 2-16.

[38] Ibid., 5-6, 5-8.

[39] Heat load is measured in terms of heating degree-days. Energy Information Administration, *Annual Energy Review 1997*, 55, 23.

per household will decline by 25% between 1996 and 2020.[40] Therefore, the reduction in space heating energy demand per household needed to meet the Kyoto Protocol levels, estimated at about 27% as described above, would have to come on top of the gains already forecast by DOE.

Industrial Direct Heat. This end-use is used for a variety of manufacturing processes. It is used principally for primary metal production such as in steel mill blast furnaces, to drive chemical reactions in chemical plants and petroleum refining, for glass and clay product manufacturing, and in food processing. To meet the Kyoto Protocol levels, industrial direct heat use 3would have to decline by about 1.5 Quads, or 18.7%, from the 1996 level, and 2.6 Quads, or 28.7% from the forecast 2008 level. The EIA expects industrial direct heat energy demand to grow about 14.7% from now to 2008, because of continued growth in the manufacturing sector. By 2010, the EIA forecast that manufacturing output will grow by about 42.5%.[41] The difference between the two growth rates is, in part, a result of expected increases in industrial process efficiency for those processes involving direct heat. In addition, manufacturing output in industries that use little or no direct heat are expected to grow faster than those that use a great deal.

To meet the Kyoto Protocol levels for industrial direct heat use, manufacturers would have to increase the efficiency of process heat equipment, increase process productivity for those goods requiring direct heat, reduce output, or undertake some combination of the three steps. An average heater efficiency increase of 23% would be required to reduce 1996 direct heat energy demand to the Kyoto target level. To maintain that level to 2008, given the current EIA forecast, would require a total heater efficiency increase of over 56% if that was the only action taken. That increase would have to come on top of efficiency increases already forecast by EIA.

To see the effect of meeting the Kyoto Protocol by reducing production, consider petroleum refining. In 1996, the United States produced about 17 million barrels per day (MMBD) of refined products. If the U.S. petroleum refining industry had to reduce its direct heat energy use to contribute its share in meeting the Kyoto Protocol level, production would have to decline by about 18% or about 3 MMBD. In 2008, the EIA forecast that U.S. refineries will produce about 19.5 MMBD.[42] In order to meet the Kyoto Protocol by reducing production, U.S. refinery output would have to drop by about 5.5 MMBD to a total of 14 MMBD. Similar analysis could be carried out for other manufacturing areas requiring direct heat in their manufacturing process.

No historical data exist specifically on direct heat energy demand so it is not possible to compare, directly, the requirements of the Kyoto Protocol, as discussed in the preceding paragraph, with past trends. It is possible, however, to determine industrial energy intensity by calculating the ratio of industrial energy demand to industrial output given in dollars.[43] From 1977 to 1989, industrial energy intensity declined by over 30%. The decline in industrial direct heat energy intensity needed to meet the Kyoto Protocol

---

[40] Energy Information Administration, *Annual Energy Outlook, 1998.* 41.

[41] Ibid., 125.

[42] Ibid., 116

[43] Council of Economic Advisors, Office of the President, *Economic Report of the President* (February 1998), 297.

from the 1996 level is about 18%. If one assumes that total industrial energy intensity and direct energy intensity track, then it would seem that meeting the Kyoto Protocol requirements is reasonable based on historical precedent. It should be noted, however, that total industrial energy intensity stopped its decline in 1989 and has risen slightly – about 5% - between 1989 and 1996. Furthermore, the 18% decline required by the Kyoto Protocol would have to come on top of the efficiency gains already forecast by EIA. It projects a decline of 17% in industrial energy intensity between 1996 and 2010.[44] Again, assuming direct heat energy intensity and total energy intensity for industry growth (or decline) at the same rate,[45] the total decline in energy intensity required to meet the Kyoto requirements while allowing industrial output to grow to levels now forecast by EIA to 2008 would be over 40%. This is well beyond any historical changes.

As for reducing output, the question is whether consumers can accommodate less production, not whether industry can produce less because of lower energy demand. In the case of oil refinery production, from 1978 to 1983, oil products supplied in the United States declined by 3.6 MMBD. This exceeds the decline from 1996 supply needed for oil refineries to meet the Kyoto Protocol if they were to choose the lower production path. Accommodation to that production decline between 1978 and 1983 was a complicated combination of fuel switching, reduced economic output and increased energy efficiency. Since 1983, however, the volume of oil products supplied has increased steadily to a current level near the 1978 peak. Furthermore, the EIA forecasts continued increases as noted above. The reduction that would be needed for oil refinery output to meet the Kyoto Protocol from the 2008 forecast level – 5.5 MMBD – is considerably greater than past levels.

Concluding Comments. The analysis presented here provides a detailed view of how energy is used in the United States. It also provides a clear picture of the contribution these end-uses make to the buildup of carbon dioxide in the earth's atmosphere. Finally, it presents a way to analyze the contributions each end-use would make to any strategy to reduce CO_2 emissions, and the implications of those strategies in terms of particular end-uses.

Obviously, end-use disaggregation could continue beyond that given in this report. For example, different types of light vehicles and different types of industrial direct heat processes exist. In particular, it was seen above that the contribution of light trucks – sport utility vehicles, vans, etc. – to the light utility vehicle fleet is growing. Data to carry out finer breakdowns, however, exist in only a few cases, and further disaggregation would be less and less precise. While current and historical data for different components of the light duty vehicle fleet exist, projections of that mix are lacking. Furthermore, it is not clear that, with the possible exception of light duty vehicles and the commercial and residential miscellaneous categories, more disaggregation would add much to

---

[44] Energy Information Administration, *Impacts of the Kyoto Protocol on U.S. Energy Markets and Economic Activity*, 53.

[45] To the degree that future gains in industrial output are disproportionately accounted for by less energy-intensive industry, this assumption becomes less valid. Nevertheless, the decline in direct heat energy intensity to meet the Kyoto Protocol is still likely to be substantial as long as total industrial output is not to suffer.

understanding how energy is used or to the analysis of the implications of reduced energy demand.

It is clear from the examples given above that the reduction in energy demand needed to reach the Kyoto Protocol targets under the assumptions made in this report would be substantially greater than previous changes in U.S. energy demand. While reductions required from 1996 levels, for the three examples considered above, appear to be comparable to those taking place in the past, when growth in energy demand that is expected between now and 2008 or 2012 is factored in, the changes required are nearly all unprecedented. Similar observations would hold for other end-uses because of the underlying sector growth now forecast.

Achieving the energy demand reductions by increases in efficiency to maintain the growth in products and services supplied by each end-use appears to require substantial gains in equipment efficiency. While for the three cases examined the gains do not appear to be impossible, they are likely to be difficult to achieve in the 12-year period and might be increasingly costly.[46]

Strategies to achieve the Kyoto Protocol levels would probably not involve efficiency gains alone, but rather would also include fuel switching and product or service substitution. The former involves substitution of energy sources that do not have any net CO_2 emissions, such as renewables or nuclear-generated electricity, for fossil fuels. The latter involves using services or products that result in lower carbon emissions than those currently used; for example, using less energy intensive materials or modes of transportation. While such substitutions are possible, they would likely take several years to implement on a scale that would contribute significantly to carbon emission reduction.

Nevertheless, substitution, particularly zero-emission energy sources, appear to be an important consideration along with increased energy efficiency in any long-term strategy to reduce carbon emissions. While it is beyond the scope of this report to consider such substitutions in detail, an example is given here to show how that might work.[47] If one-half of the coal-fired capacity projected for the nation's electricity supply for 2008 could somehow be replaced by nuclear power and/or renewables, carbon emissions in 2008 would decline by about 15% from the current forecast. That change would lower by about two-thirds the energy reduction requirements that would be needed to meet the Kyoto Protocol levels. Furthermore, all end-uses, even those that used negligible amounts of electricity, would benefit if the burden of emission reduction were apportioned to all end-uses as is done in this report. The long lead-time needed to build new power plants combined with material and personnel constraints, along with other environmental and regulatory issues, however, would likely preclude a substitution of that magnitude within 10 years. Over a longer period, such substitution is probably more feasible.

Carbon emission reduction to meet the Kyoto Protocol levels for the 2008-2012 period by reducing energy demand for current end-uses appears to be a substantial

---

[46] For another view on this issue see Office of Energy Efficiency and Renewable Energy, Department of Energy, *Scenarios of U.S. Carbon Reductions: Potential Impacts of Energy Technologies by 2010 and Beyond.*

[47] The model built to perform the analysis presented above can also be used to calculate the effects of fuel substitution.

undertaking as seen in the above analysis. If apportioned to all end-uses, each would be affected significantly by 2008 given the reduction requirements and the currently forecast growth in that end-use. If all or a major portion of the energy demand reduction were a result of a lower level of service from that end-use rather than greater energy efficiency, consumers of those end-uses would likely be substantially affected.

APPENDIX: DETAILED DATA TABLES

The following are the detailed data tables showing for each end-use the actual and forecast energy demand and carbon emissions for 1996, 2008, and 2012, the carbon emission levels and resultant energy demand levels needed to reach the Kyoto Protocol levels of a 7% reduction from the 1990 levels, and the resultant changes from 1996, 2008, and 2012.

Energy Demand and Carbon Emissions – 1996 (Quads and Millions of Tons)

End-use	Sector	Symbol	Elect	Resid	Dist	NG	LPG	Coal	Bio	Gasol	Jet	Total	Carbon
Light Duty Veh	TRANS	LDT			0.05					13.91		13.96	270.57
Direct Heat	IND	DHI	1.05	1.71	0.03	3.82	0.11	1.26				7.98	143.62
Space Ht	RES	SHR	1.74		0.88	3.76	0.31	0.05	0.61			7.35	106.45
Trucks	TRANS	FTT			3.47	0.01				1.12		4.60	91.08
Steam	IND	StI	0.09	0.32	0.04	3.38	0.08	1.03	1.83			6.77	85.78
Machine Drive	IND	MDI	4.93		0.02	0.10						5.50	80.90
Mscl	COM	MC	3.48		0.19	1.31						4.98	78.61
Air	TRANS	ArT								0.05	3.27	3.32	64.18
Appliances	RES	ApR	3.74			0.21	0.03					3.98	63.55
Lighting	COM	LC	3.71									3.71	59.49
Const & Ag	IND	CAI	1.74		0.90	0.26	0.10			0.15		3.15	54.23
Miscl	RES	MR	2.52			0.09	0.01					2.62	41.82
Water Heat	RES	WHR	1.16		0.09	1.32	0.07					2.64	40.71
Mining	IND	MnI	0.79		0.18	1.11	0.19	0.06		0.02		2.38	38.12
VAC	COM	ACC	2.20	0.02		0.02						2.22	35.64
Space Heat	COM	SHC	0.39		0.23	1.34	0.08	0.08				2.12	33.60
Marine	TRANS	MaT		0.44	0.96							1.40	28.61
A/C	RES	ACR	1.65									1.65	26.38
HVAC	IND	HAI	0.85			0.36						1.21	18.84
Lighting	RES	LR	1.10									1.10	17.59
Electrolysis	IND	ELI	1.02									1.02	16.36
Water Heat	COM	WHC	0.55		0.05	0.45						1.05	16.30
Rail	TRANS	RT	0.19	0.46								0.65	12.99
Lighting	IND	LI	0.75									0.75	12.03
Cook & Ref	COM	CRC	0.55			0.18						0.73	11.40
Pipeline	TRANS	PT				0.73						0.73	10.56
Elec Gen	IND	ELI				0.35						0.35	5.06
Totals			34.20	2.95	7.09	18.80	0.98	2.48	2.44	15.25	3.27	87.47	1464.46

Energy Use and Carbon Emissions – 2008 (Quads and Millions of Tons)

End-use	Sector	Symbol	Elect	Resid	Dist	NG	LPG	Coal	Bio	Gasol	Jet	Total	Carbon
Light Duty Veh	TRANS	LDT			0.09	0.10				16.98		17.17	332.28
Direct Heat	IND	DHI	1.30	2.00	0.04	4.32	0.12	1.33				9.10	163.72
Frgt Trucks	TRANS	FTT			5.01					0.81		5.82	115.60
Miscl	COM	MC	4.94	0.12	0.18	1.57	0.09	0.09				6.98	114.12
Space Ht	RES	SHR	1.91		0.75	3.83	0.34	0.05	0.64			7.51	108.92
Air	TRANS	ArT								0.04	4.93	4.97	95.98
Machine Drive	IND	MDI	6.09		0.02	0.11						6.23	102.69
Steam	IND	StI		0.37	0.05	3.82	0.09	1.08	2.29			7.82	95.45
Miscl	RES	MR	4.50			0.10	0.01					4.61	75.93
Lighting	COM	LC	3.71									3.71	61.20
Appliances	RES	ApR	3.37			0.23	0.04					3.63	59.52
Const & Ag	IND	CAI	1.28		0.96	0.32	0.11					2.67	46.82
Mining	IND	MnI	0.98	0.02	0.22	1.26	0.11	0.06		0.03		2.67	43.10
Water Heat	RES	WHR	1.13		0.10	1.41	0.10					2.74	42.78
Marine	TRANS	MaT		0.77	0.92							1.68	34.74
VAC	COM	ACC	2.18			0.02						2.20	36.33
Space Heat	COM	SHC	0.42		0.18	1.38						1.98	30.55
A/C	RES	ACR	1.63									1.63	26.88
HVAC	IND	HAI	1.05			0.41						1.46	23.23
Lighting	RES	LR	1.26									1.26	20.86
Electrolysis	IND	ELI	1.15									1.15	19.06
Rail	TRANS	RT	0.42	0.44								0.86	16.35
Water Heat	COM	WHC	0.45		0.05	0.51						1.02	15.89
Pipeline	TRANS	PT				0.81						0.81	11.68
Cook & Ref	COM	CRC	0.57			0.22						0.79	12.56
Lighting	IND	LI	0.57									0.57	9.37
Elec Gen	IND	ELI				0.40						0.40	5.73
Totals			39.04	3.72	8.56	20.80	1.01	2.61	2.93	17.85	4.93	101.45	1721.33

Energy Demand and Carbon Emissions – 2012 (Quads and Millions of Tons)

End-use	Sector	Symbol	Elect	Resid	Dist	NG	LPG	Coal	Bio	Gasol	Jet	Total	Carbon
Light Duty Veh	TRANS	LDT			0.09	0.12				18.20		18.42	356.35
Direct Heat	IND	DHI	1.33	1.98	0.04	4.41	0.13	1.33				9.21	165.24
Frgt Trucks	TRANS	FTT			5.20					0.84		6.04	120.01
Miscl	COM	MC	5.07	0.12	0.18	1.61	0.09	0.09				7.15	116.93
Space Ht	RES	SHR	1.99		0.73	3.92	0.34	0.05	0.65			7.68	111.22
Air	TRANS	ArT								0.04	5.45	5.49	106.21
Machine Drive	IND	MDI	6.26		0.03	0.12						6.40	105.47
Steam	IND	StI	0.11	0.37	0.05	3.90	0.09	1.08	2.38			7.99	96.68
Miscl	RES	MR	4.68			0.10	0.01					4.79	78.89
Lighting	COM	LC	3.80									3.80	62.78
Appliances	RES	ApR	3.50			0.23	0.04					3.77	61.79
Const & Ag	IND	CAI	1.32		1.01	0.32	0.11					2.76	48.43
Mining	IND	MnI	1.00	0.02	0.23	1.28	0.11	0.06		0.03		2.74	44.20
Water Heat	RES	WHR	1.17		0.10	1.45	0.10					2.82	44.03
Marine	TRANS	MaT		0.85	0.95							1.80	37.19
VAC	COM	ACC	2.24			0.02						2.26	37.26
Space Heat	COM	SHC	0.43		0.18	1.41						2.03	31.19
A/C	RES	ACR	1.69									1.69	27.94
HVAC	IND	HAI	1.08			0.42						1.49	23.83
Lighting	RES	LR	1.31									1.31	21.68
Electrolysis	IND	ELI	1.19									1.19	19.58
Rail	TRANS	RT	0.51	0.48								0.99	18.80
Water Heat	COM	WHC	0.46		0.05	0.53						1.04	16.25
Pipeline	TRANS	PT				0.98						0.98	14.21
Cook & Ref	COM	CRC	0.59			0.22						0.81	12.88
Lighting	IND	LI	0.58									0.58	9.62
Elec Gen	IND	ELI				0.40						0.40	5.85
Totals			40.32	3.82	8.82	21.46	1.03	2.62	3.02	19.11	5.45	105.66	1794.51

End-Use Energy Demand and Carbon Emissions (Quadrillions of BTUs and Millions of Metric Tons)

End-use	2008 Forecast		1996 Actual		2008 Kyoto Target		Energy Reduction		Carbon Reductions	
	Total	Carbon	Energy	Carbon	Energy	Carbon	From 2008	From 1996	From 2008	From 1996
LDT	17.17	332.28	13.96	270.57	12.23	236.78	4.93	1.73	95.51	33.79
DHI	9.10	163.72	7.98	143.62	6.49	116.66	2.62	1.49	47.06	26.96
FTT	5.82	115.60	4.60	91.08	4.15	82.37	1.67	0.45	33.23	8.70
MC	6.98	114.12	4.98	78.61	4.97	81.32	2.01	0.01	32.80	-2.71
SHR	6.88	108.92	6.74	106.45	4.90	77.61	1.98	1.84	31.31	28.83
ArT	4.97	95.98	3.32	64.18	3.54	68.39	1.43	-0.22	27.59	-4.21
MDI	6.23	102.69	5.05	80.90	4.44	73.17	1.79	0.61	29.52	7.73
StI	5.53	95.45	4.94	85.78	3.94	68.02	1.59	1.00	27.44	17.77
MR	4.61	75.93	2.62	41.82	3.29	54.11	1.33	-0.67	21.82	-12.29
LC	3.71	61.20	3.71	59.49	2.64	43.61	1.07	1.07	17.59	15.88
ApR	3.63	59.52	3.98	63.55	2.59	42.41	1.04	1.39	17.11	21.14
CAI	2.67	46.82	3.15	54.23	1.90	33.36	0.77	1.25	13.46	20.86
MnI	2.67	43.10	2.38	38.12	1.90	30.71	0.77	0.48	12.39	7.41
WHR	2.74	42.78	2.64	40.71	1.95	30.48	0.79	0.69	12.30	10.22
MaT	1.68	34.74	1.40	28.61	1.20	24.76	0.48	0.20	9.99	3.85
ACC	2.20	36.33	2.22	35.64	1.57	25.88	0.63	0.65	10.44	9.75
SHC	1.98	30.55	2.12	33.60	1.41	21.77	0.57	0.70	8.78	11.83
ACR	1.63	26.88	1.65	26.38	1.16	19.16	0.47	0.48	7.73	7.23
HAI	1.46	23.23	1.21	18.84	1.04	16.55	0.42	0.17	6.68	2.29
LR	1.26	20.86	1.10	17.59	0.90	14.87	0.36	0.20	6.00	2.72
EI	1.15	19.06	1.02	16.36	0.82	13.58	0.33	0.20	5.48	2.78
RT	0.86	16.35	0.65	12.99	0.61	11.65	0.25	0.04	4.70	1.34
WHC	1.02	15.89	1.05	16.30	0.72	11.32	0.29	0.32	4.57	4.98
PT	0.81	11.68	0.73	10.56	0.58	8.33	0.23	0.15	3.36	2.24
CRC	0.79	12.56	0.73	11.40	0.56	8.95	0.23	0.17	3.61	2.45
LI	0.57	9.37	0.75	12.03	0.40	6.68	0.16	0.35	2.69	5.35
EGI	0.40	5.73	0.35	5.06	0.28	4.08	0.11	0.07	1.65	0.98
Totals	98.52	1721.33	85.03	1464.46	70.20	1226.59	28.32	14.83	494.75	237.88

End-Use Energy Demand and Carbon Emissions (Quadrillions of BTUs and Millions of Metric Tons)

End-use	2012 Forecast		2008 Kyoto Target		Reductions	
	Energy	Carbon	Energy	Carbon	Energy	Carbon
LDT	18.42	356.35	12.23	236.78	6.18	119.57
DHI	9.21	165.35	6.49	116.66	2.73	48.58
FTT	6.04	120.01	4.15	82.37	1.89	37.63
MC	7.15	116.93	4.97	81.32	2.18	35.60
SHR	7.03	111.22	4.90	77.61	2.13	33.61
ArT	5.49	106.21	3.54	68.39	1.96	37.82
MDI	6.40	105.47	4.44	73.17	1.96	32.29
StI	5.61	96.68	3.94	68.02	1.67	28.67
MR	4.79	78.89	3.29	54.11	1.51	24.79
LC	3.80	62.78	2.64	43.61	1.16	19.17
ApR	3.77	61.79	2.59	42.41	1.18	19.38
CAI	2.76	48.43	1.90	33.36	0.86	15.06
MnI	2.74	44.20	1.90	30.71	0.84	13.49
WHR	2.82	44.03	1.95	30.48	0.87	13.55
MaT	1.80	37.19	1.20	24.76	0.60	12.44
ACC	2.26	37.26	1.57	25.88	0.69	11.38
SHC	2.03	31.19	1.41	21.77	0.61	9.42
ACR	1.69	27.94	1.16	19.16	0.53	8.79
HAI	1.49	23.83	1.04	16.55	0.46	7.27
LR	1.31	21.68	0.90	14.87	0.41	6.82
EI	1.19	19.58	0.82	13.58	0.36	5.99
RT	0.99	18.80	0.61	11.65	0.38	7.15
WHC	1.04	16.25	0.72	11.32	0.32	4.93
PT	0.98	14.21	0.58	8.33	0.41	5.88
CRC	0.81	12.88	0.56	8.95	0.25	3.93
LI	0.58	9.62	0.40	6.68	0.18	2.95
EGI	0.40	5.85	0.28	4.08	0.12	1.77
Totals	102.64	1794.51	70.20	1226.59	32.43	567.93

INDEX

D

V

W

Y